10th International Conference on Power, Energy and Electrical Engineering (CPEEE 2020)

AA003614

E3S Web of Conferences Volume 182 (2020)

Tokyo, Japan
19 - 21 June 2020

Editors:

E. Sutheerasak
M.L. Kolhe
M. Popescu
H. Li
J. Cale
S. Mirsaeidi
K. Ogura

ISBN: 978-1-7138-1412-2

Printed from e-media with permission by:

Curran Associates, Inc.
57 Morehouse Lane
Red Hook, NY 12571

proceedings.com

Some format issues inherent in the e-media version may also appear in this print version.

No changes have been made to the content of these proceedings. There may be changes to pagination and minor adjustments for aesthetics.

Printed with permission by Curran Associates, Inc. (2021)

For additional information, please contact EDP Sciences – Web of Conferences at the address below.

EDP Sciences – Web of Conferences
17, Avenue du Hoggar
Parc d'Activité de Courtabœuf
BP 112
F-91944 Les Ulis Cedex A
France

Phone: +33 (0) 1 69 18 75 75
Fax: +33 (0) 1 69 28 84 91

contact-edps@webofconferences.org

Additional copies of this publication are available from:

Curran Associates, Inc.
57 Morehouse Lane
Red Hook, NY 12571 USA
Phone: 845-758-0400
Fax: 845-758-2633
Email: curran@proceedings.com
Web: www.proceedings.com

TABLE OF CONTENTS

The 10th International Conference on Power, Energy and Electrical Engineering

CPEEE 2020, June 19-21, 2020

Editorial

The 2020 10th International Conference on Power, Energy and Electrical Engineering (CPEEE 2020) is organized as a fully virtual conference during 19-21 June 2020. The CPEEE 2020 is technical sponsored by the Tokyo University of Technology (Japan) and supported by Texas Tech University (USA), Kyushu Sangyo University (Japan), and Pusan National University (Korea).

The CPEEE 2020 is aimed at providing an excellent avenue for academicians, students, researchers, professionals, engineers, and scientists from academia and industry to share their research findings and building network for further collaborative research in their respective areas of power, energy and electrical engineering.

The Technical Programme Committee of CPEEE 2020 has assembled an excellent programme, which has included keynote presentations from recognized academicians: Prof. Ryuichi Yokoyama, Waseda University (Japan); Prof. C.Y. (Tony) Chung, University of Saskatchewan, (Canada); Prof. Ryuji Matsuhashi, The University of Tokyo (Japan), Prof. Mohan Lal Kolhe, University of Agder (Norway); Invited Speaker: Prof. Koki Ogura, Kyushu Sangyo University, (Japan); and technical papers presentation have organized in 4 parallel sessions.

This CPEEE 2020 encompasses advancement in sustainable electrical power and energy system with focus on sustainable energy system engineering, integration of renewable energy technologies, electro-chemical energy storage, energy conversion technologies, etc. The key aspects of electro-chemical energy storage including hydrogen production and utilization processes are also covered. The proceeding of CPEEE 2020 consists of 23 selected peer reviewed papers, which were submitted to the conference from the universities, research institutes and industries. All papers are peer-reviewed by the CPEEE 2020 technical committee members and the renowned international reviewers.

As the editor of this special issue of the CPEEE 2020 proceeding, I wish to express my sincere gratitude to the keynote speakers, invited speakers, authors, reviewers, program and technical committee members, who have significantly contributed in organization of the conference and publication of this special issue. My special thanks to E3S Web of Conferences for providing the platform to publish the peer-reviewed conference papers. With high-quality international submissions and knowledge disseminations in the CPEEE 2020, I am assured that the CPEEE will become an important conference in multi-discipline domains of power and energy systems of sustainable electrical engineering.

At last, I am confident that the contents of this special volume will be useful for the researchers and practitioners in developing and applying emerging technologies in the field of power, energy and electrical engineering.

Professor Mohan Lal Kolhe
Editor of CPEEE 2020 Proceeding and Conference Program Chair of CPEEE 2020,
Professor in Smart Grid and Renewable Energy
Faculty of Engineering and Science,
University of Agder,
PO Box 422, NO 4604, Kristiansand, Norway

Conference Committees

Conference Advisory Committees
- Prof. C.Y. (Tony) Chung, IEEE Fellow, University of Saskatchewan, Canada
- Prof. Lalit Kumar Goel, IEEE Fellow, Nanyang Technological University, Singapore
- Prof. Muhammad Arfin Khan Lodhi, Texas Tech University, USA

Conference Chair
- Prof. Ryuichi Yokoyama, IEEE Fellow, Waseda University, Japan

Conference Local Chairs
- Prof. Yasuhiro Noro, Kogakuin University, Japan
- Prof. Junichi Arai, Kogakuin University, Japan

Conference Program Chairs
- Prof. Jinhua She, Tokyo University of Technology, Japan
- Prof. Koki Ogura, Kyushu Sangyo University, Japan
- Prof. Hee-Je Kim, Pusan National University, Korea
- Prof. Mohan Lal Kolhe, Universitetet of Agder, Norway

Conference Technical Committees
- Prof. Ichijo Hodaka, University of Miyazaki, Japan
- Prof. José Fernando Alves da Silva, Instituto Superior Técnico, Portugal
- Prof. Piotr Chrzan, Politechnika Gdanska, Poland
- Prof. Tarlochan Sidhu, University of Ontario Institute of Technology, Canada
- Assoc. Prof. Amr Elnady, University of Sharjah, UAE
- Prof. Michael Negnevitsky, University of Tasmania, Australia
- Prof. Dr. Pu Li, Technische Universitat Ilmenau, Germany
- Prof. Dr. Lingai Luo, Polytech' Nantes - Université de Nantes, France
- Prof. Dr. Talal Yusaf, Federation University Australia, Australia
- Prof. Lieven Vandevelde, Ghent University, Belguim
- Dr. Sadegh Azizi, University of Leeds, UK
- Assoc. Prof. James Cale, Colorado State University, USA
- Dr. Mattia Ricco, University of Bologna, Italy
- Assoc. Prof. Angela Russo, Politecnico di Torino, Dipartimento Energia "Galileo Ferraris", Italy
- Assoc. Prof. Hua Li, Texas A&M University-Kingsville, USA
- Dr. Chunhua Liu, City University of Hong Kong, Hong Kong
- Prof. Dr. Mihaela Popescu, University of Craiova, Romania
- Asst. Prof. Ekkachai Sutheerasak, Burapha University, Thailand
- Dr. Smith Eiamsa-ard, Mahanakorn University of Technology, Thailand
- Assoc. Prof. Saleh Mohammed Saleh Shalaby, Tanta University, Egypt
- Assoc. Prof. Sohrab Mirsaeidi, Beijing Jiaotong University, China
- Assoc. Prof. Pooya Davari, Aalborg University, Denmark
- Prof. Diego Bellan, Politecnico di Milano, Italy
- Asst. Prof. Dr. Hassan Ali, University of Newcastle, Australia
- Assoc. Prof. Chong Wen Tong, University of Malaya, Malaysia
- Assoc. Prof. Pierluigi Siano, University of Salerno, Italy
- Assoc. Prof. Qiuwei Wu, Technical University of Denmark, Denmark
- Dr. Chin Hsia, National Central University, Taiwan
- Dr. Xinggang Yan, University of Kent, UK
- Dr. Wenxian Yang, Newcastle University, UK
- Dr. Farhad Shahnia, Murdoch University, Australia

edp sciences

E3S Web of Conferences

Statement of Peer review

In submitting conference proceedings to E3S Web of Conferences, I certify to the Publisher that I adhere to the **Policy on Publishing Integrity** of the journal in order to safeguard good scientific practice in publishing.

1. All articles have been subjected to peer review administered by the proceedings editors.
2. Reviews have been conducted by expert referees, who have been requested to provide unbiased and constructive comments aimed, whenever possible, at improving the work.
3. Proceedings editors have taken all reasonable steps to ensure the quality of the materials they publish and their decision to accept or reject a paper for publication has been based only on the merits of the work and the relevance to the journal.

Title, date and place of the conference

2020 10th International Conference on Power, Energy and Electrical Engineering (CPEEE 2020)
Tokyo, Japan
June 19-21, 2020

Proceedings editor(s):

Asst. Prof. Ekkachai Sutheerasak, Burapha University, Thailand

Date and editor's signature June 15, 2020

Statement of Peer review

In submitting conference proceedings to E3S Web of Conferences, I certify to the Publisher that I adhere to the **Policy on Publishing Integrity** of the journal in order to safeguard good scientific practice in publishing.

1. All articles have been subjected to peer review administered by the proceedings editors.
2. Reviews have been conducted by expert referees, who have been requested to provide unbiased and constructive comments aimed, whenever possible, at improving the work.
3. Proceedings editors have taken all reasonable steps to ensure the quality of the materials they publish and their decision to accept or reject a paper for publication has been based only on the merits of the work and the relevance to the journal.

Title, date and place of the conference

2020 10th International Conference on Power, Energy and Electrical Engineering (CPEEE 2020)
Tokyo, Japan
June 19-21, 2020

Proceedings editor(s):

Prof. Mohan Lal Kolhe, Universitetet of Agder, Norway

15.06.2020

Date and editor's signature

Statement of Peer review

In submitting conference proceedings to E3S Web of Conferences, I certify to the Publisher that I adhere to the **Policy on Publishing Integrity** of the journal in order to safeguard good scientific practice in publishing.

1. All articles have been subjected to peer review administered by the proceedings editors.
2. Reviews have been conducted by expert referees, who have been requested to provide unbiased and constructive comments aimed, whenever possible, at improving the work.
3. Proceedings editors have taken all reasonable steps to ensure the quality of the materials they publish and their decision to accept or reject a paper for publication has been based only on the merits of the work and the relevance to the journal.

Title, date and place of the conference

2020 10th International Conference on Power, Energy and Electrical Engineering (CPEEE 2020)
Tokyo, Japan
June 19-21, 2020

Proceedings editor(s):

Prof. Dr. Mihaela Popescu, University of Craiova, Romania

Date and editor's signature 15 June 2020

17, avenue du Hoggar - PA de Courtabœuf – BP 112 - 91944 Les Ulis Cedex A (France)
Tél. : 33 (0)1 69 18 75 75 - Fax : 33(0)1 69 07 45 17 – www.edpsciences.org

Statement of Peer review

In submitting conference proceedings to E3S Web of Conferences, I certify to the Publisher that I adhere to the **Policy on Publishing Integrity** of the journal in order to safeguard good scientific practice in publishing.

1. All articles have been subjected to peer review administered by the proceedings editors.
2. Reviews have been conducted by expert referees, who have been requested to provide unbiased and constructive comments aimed, whenever possible, at improving the work.
3. Proceedings editors have taken all reasonable steps to ensure the quality of the materials they publish and their decision to accept or reject a paper for publication has been based only on the merits of the work and the relevance to the journal.

Title, date and place of the conference

2020 10th International Conference on Power, Energy and Electrical Engineering (CPEEE 2020)
Tokyo, Japan
June 19-21, 2020

Proceedings editor(s):

Assoc. Prof. Hua Li, Texas A&M University-Kingsville, USA

6-15-2020

Date and editor's signature

Statement of Peer review

In submitting conference proceedings to E3S Web of Conferences, I certify to the Publisher that I adhere to the **Policy on Publishing Integrity** of the journal in order to safeguard good scientific practice in publishing.

1. All articles have been subjected to peer review administered by the proceedings editors.
2. Reviews have been conducted by expert referees, who have been requested to provide unbiased and constructive comments aimed, whenever possible, at improving the work.
3. Proceedings editors have taken all reasonable steps to ensure the quality of the materials they publish and their decision to accept or reject a paper for publication has been based only on the merits of the work and the relevance to the journal.

Title, date and place of the conference

2020 10th International Conference on Power, Energy and Electrical Engineering (CPEEE 2020)
Tokyo, Japan
June 19-21, 2020

Proceedings editor(s):

Assoc. Prof. James Cale, Colorado State University, USA

Date and editor's signature 15 JUNE 20

Statement of Peer review

In submitting conference proceedings to E3S Web of Conferences, I certify to the Publisher that I adhere to the **Policy on Publishing Integrity** of the journal in order to safeguard good scientific practice in publishing.

1. All articles have been subjected to peer review administered by the proceedings editors.
2. Reviews have been conducted by expert referees, who have been requested to provide unbiased and constructive comments aimed, whenever possible, at improving the work.
3. Proceedings editors have taken all reasonable steps to ensure the quality of the materials they publish and their decision to accept or reject a paper for publication has been based only on the merits of the work and the relevance to the journal.

Title, date and place of the conference

2020 10th International Conference on Power, Energy and Electrical Engineering (CPEEE 2020)
Tokyo, Japan
June 19-21, 2020

Proceedings editor(s):

Assoc. Prof. Sohrab Mirsaeidi, Beijing Jiaotong University, China

Date and editor's signature June 16, 2020

Statement of Peer review

In submitting conference proceedings to E3S Web of Conferences, I certify to the Publisher that I adhere to the **Policy on Publishing Integrity** of the journal in order to safeguard good scientific practice in publishing.

1. All articles have been subjected to peer review administered by the proceedings editors.
2. Reviews have been conducted by expert referees, who have been requested to provide unbiased and constructive comments aimed, whenever possible, at improving the work.
3. Proceedings editors have taken all reasonable steps to ensure the quality of the materials they publish and their decision to accept or reject a paper for publication has been based only on the merits of the work and the relevance to the journal.

Title, date and place of the conference

2020 10th International Conference on Power, Energy and Electrical Engineering (CPEEE 2020)
Tokyo, Japan
June 19-21, 2020

Proceedings editor(s):

Prof. Koki Ogura, Kyushu Sangyo University, Japan

Date and editor's signature

June 17, 2020

Applicability of Low Frequency Transmission Technology in Offshore Wind Power Transmission Scene

Jian Yang[1], *Jing* Ge[2,*], *Zuofu* Jiang[3], *Chi* Zhang[1], *Jianhua* Lv[1], *Guode* Ying[1], *Haohua* Wu[4]

[1] State Grid Taizhou Electric Power Supply Company, 31800, China
[2] State Key Laboratory of Advanced Transmission Technology (Global Energy Interconnection Researching Institute Co., Ltd.), State Grid Office Area, North District of Future Science and Technology City, Changping District, Beijing, 102200, China
[3] Taizhou Hongchuang Electricity Group Limited Company, 318000, China
[4] Taizhou Hongda Electric Construction Co., Ltd. Taizhou Economic Development Zone Operation And Mintenance Branch, 318000, China

Abstract: This paper investigates the worldwide development status of wind power and the main opinions and advantages of offshore wind power development in the industry, and summarizes that the development of offshore wind power will be in the direction of large capacity and distant sea area. At the same time, the distribution of wind energy resources in China's coastal areas is investigated, and combined with the constraints of offshore wind energy development in China, it is concluded that offshore wind power development in China will also develop to distant sea areas. Secondly, three transmission modes including power frequency AC, flexible HVDC and low frequency AC are briefly described, and the technical economy of the three modes is analyzed with typical cases. In addition, the technical feasibility of 400 MW offshore wind power Grid-connected through low-frequency transmission technology is verified by system model building and simulation. Thirdly, combined with the development trend of offshore wind power, current research and manufacturing level, the opportunities and challenges of the application of low frequency transmission technology are analyzed. Finally, the main points of this paper are summarized. It is preliminarily clear that the low-frequency transmission mode has better technical and economic performance within a certain transmission distance, and can be used as a beneficial supplement to the conventional power frequency and flexible HVDC transmission mode.

1 Introduction

In recent years, wind power has developed rapidly. According to the Global Wind Power Development Report 2017[1] issued by the Global Wind Energy Council, more than 90 countries in the world have carried out wind power commercialization business, of which 9 countries have installed more than 10 GW wind power and 30 countries (Europe, Asia, North America, Latin America and Africa) have installed more than 1 GW wind power. By the end of December 2017, the total installed capacity of wind power in the world is 539.1 GW. The report predicts that the total installed capacity of wind power will reach 840 GW by the end of 2022. In its Global Wind Power Development Report 2016[2], the Global Wind Energy Council predicted that in 2020 and 2030, the installed wind power in the world would reach 560 GW and 1100 GW, respectively, and the generating capacity would reach 1.4 trillion kWh and 2.8 trillion kWh, respectively.

In addition to land wind power, offshore wind power is developing rapidly and has a broad market. In 2016, the International Renewable Energy Agency (IRENA) published a report entitled Innovation Outlook: Offshore Wind[3], which assessed the future trend of the factors related to the rapid development of the offshore wind power industry. If the global economy has been developing towards a non-carbon direction, wind power will become the main power supply by 2030. According to the International Renewable Energy Agency, the total installation of offshore wind power will reach 100 GW by 2030.

The development of offshore wind power industry now covers not only the Nordic region, but also North America, East Asia, India and other regions. In the foreseeable future, although Europe will continue to lead the offshore wind power market, the current trend shows that the global offshore wind power is likely to start to develop rapidly in the next few years.

Compared with onshore wind energy, offshore wind energy has the following advantages[4-6]:

(1) The offshore wind resources are abundant, and the wind speed is relatively stable, which is larger than the onshore wind power generation. Usually, the sea surface wind speed of 10 km offshore is about 25% larger than that of the coastal land, and is less affected by the environment. Wind resources can be used three times as much as that of the land. It is more conducive to the collection and conversion of wind energy resources. It

* Corresponding author: gjsonia@sina.com

has no great interference to human activities and reduces the cost of investment.

(2) The surface roughness of sea water is low, the friction force at sea level is small, and the wind speed varies little with the height. It does not need a high tower to reduce the cost of wind turbines.

(3) The turbulence intensity of offshore wind is low, and the fatigue load of wind action on wind turbines is reduced, thus prolonging the service life of wind turbines.

(4) Wind power technology has been relatively mature, with the most large-scale development and commercial development. Some European countries' offshore wind power has entered the commercial stage after 2001, and the technology will be gradually improved.

(5) The frequency of wind speed available for wind energy development in most of the world's sea areas is basically more than 60%. That is to say, most of the world's sea areas can start wind power generation at least half of the year, while solar energy and other limited by daytime, the available time will not exceed 50%.

Therefore, large-scale offshore wind power development has become the development direction of new energy utilization at home and abroad, and will develop towards large capacity, Far-sea and deep-water areas. The offshore wind power transmission technology is particularly important. Traditional power frequency transmission is only suitable for wind farms with small capacity and close offshore distance. Flexible DC technology can meet the needs of large capacity transmission, but its construction and maintenance costs are high. Low-frequency transmission technology can effectively improve the current carrying capacity of submarine cable and reduce terminal voltage fluctuation by reducing transmission frequency. It has good economy in a certain transmission distance. This paper will compare three kinds of offshore wind power transmission modes from technical and economic aspects, and give the applicable scope of each transmission mode.

2 Development status and delivery technology analysis of offshore wind power

2.1 Recent development of offshore wind power in china

In 2010, the National Climate Center simulated the distribution of wind energy resources at 70 m altitude in China's offshore waters as shown in Table 1. China's offshore wind energy resources are abundant, with the East China Sea being the most abundant of the four major sea areas, followed by the Bohai Sea, the South China Sea and the Yellow Sea. Among all the sea areas, the Taiwan Strait is the most abundant in wind energy resources, with an average wind speed of over 8.5 m/s and over 9 m/s in some areas; followed by Zhejiang coast, eastern coast of Guangdong and Liaodong Bay of Bohai Sea, with an average wind speed of over 8 m/s; the average wind speed of other sea areas is mostly above 7.5 m/s; the wind energy resources in northern Beibu Gulf and central Yellow Sea are relatively poor, with an average wind speed of about 7 m/s. According to the wind power report of 2014, in the coastal areas of China, the wind energy reserves are 200 GW in the areas of 5-25 meters deep and 50 meters high, and 500 GW in the areas of 5-50 meters deep and 70 meters high[7].

From the analysis of Table 1, we can see that the sea area between Liaodong Peninsula and Korean Peninsula, the east of Bohai Bay and Leizhou Peninsula in Guangdong Province, and the sea area from Yangtze River estuary to Hangzhou Bay have abundant energy reserves. In addition, in recent years, China has put into operation offshore wind farms, but offshore areas are restricted by military, fishery, environmental protection and other factors, so the development of offshore wind energy in China will also be developed to distant areas.

Table 1 Distribution of offshore wind power resources in China

Areas	average annual wind speed (m/s)	average wind power density (w/m^2)
Coastal Yellow Sea	5m: 7.0-7.5	250-450
Shandong Peninsula and Laizhou Bay	25m: 8.0	
Weihai		
Coastal area of Eastern Liaodong Peninsula	5m: 7.0-7.5	300-500
Dalian, Southern End of Liaodong Peninsula	25m: 8.0	
Liaodong Bay		
Bohai Bay		
Offshore Jiangsu	6.0-6.5	250-400
North-central Jiangsu		
Southern Jiangsu		300-400
Zhejiang Offshore	50m: 8.0	250-450
Hangzhou Bay, Zhejiang		
The rest of Zhejiang sea area		
Fujian and northeastern Guangdong	50m: 8.0	300-800
The Eastern Sea Area of Leizhou Peninsula, Guangdong		300-450
Hainan	6.5-7.5	250-400
Offshore of Guangxi	25m: 6.0-6.5 50m: 6.5-7.5	250-450

According to the distribution of wind energy resources in China, the relevant departments of China have issued reports on offshore wind power development planning. In the letter of A Reply to the Planning of Offshore Wind Power in Fujian Province[8], it was pointed out that the State Energy Administration agreed that the total scale of offshore wind power planning in Fujian Province was 13.3 million kilowatts, including 17 wind farms in Fuzhou, Zhangzhou, Putian, Ningde and Pingtan. By the end of 2020, the installed scale of offshore wind power in Fujian Province would reach more than 2 million kilowatts and by the end of 2030 to more than 5 million kilowatts. Among the 17 wind farms planned in Fujian Province, the scale of each area varies from 100 MW to 500 MW, and the voltage level is 220 kV. The theoretical water depth developed recently is below 20 meters, and the theoretical water depth in the medium or long term is between 25 and 60 meters. The transmission distance of different wind fields varies greatly, the nearest is 1.6 km, and the farthest is 70 km.

The Development Plan of Offshore Wind Power in Guangdong Province(2017-2030) (Revision)[9] planned the development of offshore wind power from 2017 to 2030. The offshore distance is not less than 10 kilometers and the water depth is less than 50 meters. It is estimated that by the end of 2020, the installed capacity of offshore wind power will exceed 12 million kilowatts, and by the end of 2030, the installed capacity of offshore wind power will be about 30 million kilowatts. The province plans 23 offshore wind farm sites, 15 offshore wind farm sites in shallow waters (within 35 meters of water depth), 8 offshore wind farm sites in deep waters (35-50 meters of water depth), the nearest 10 km offshore distance and the farthest 170 km offshore. The installed capacity of a single wind farm varies from 100MW to 900MW.

China is rich in offshore wind power resources in Fujian, Guangdong and Guangxi, and the development of offshore wind power over medium and long distances has been included in the plan, which will be the main trend of offshore wind power development in the future.

2.2 Analysis of existing offshore wind power transmission technology

In recent years, offshore wind power development in China's offshore areas has entered a stage of large-scale and commercial development. There are two traditional modes of offshore wind power transmission and grid-connected.

2.2.1 High voltage ac grid connection technology

At present, high-voltage AC transmission is the main way of offshore wind farms, such as the British Matrix offshore wind farm, the East China Sea such as the East China Sea wind farm and so on. High voltage AC transmission needs to change the output voltage level of wind farm and three-phase AC submarine cable transmission[10]. Therefore, large capacity transformer must be used to change the output voltage level of wind farm. The structure of high voltage AC transmission system is shown in Figure 1. High-voltage AC technology is relatively mature, but such wind farms and integrated power grid coupling, will have interaction. Due to the use of cable transmission, the capacitive reactance and inductance of lines, especially the relatively large capacitance of submarine cables to the ground, need reactive power compensation. Therefore, high voltage AC transmission is only suitable for short-distance wind power transmission[11-12].

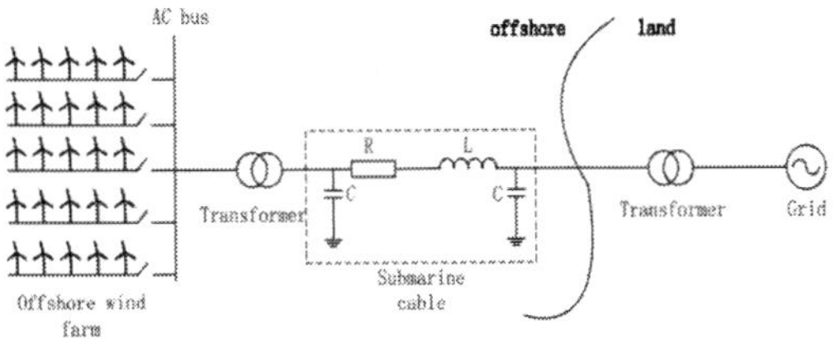

Fig. 1. High voltage AC transmission

2.2.2 Flexible DC grid-connected technology

The structure of flexible DC transmission system is shown in Fig. 2. Because there is no reactive power compensation problem in DC, there is no limit on transmission distance, so it is more suitable for long-distance power transmission[12-14]. In addition, the general high voltage DC cable is also less than three-phase AC, the cost of the cable is relatively reduced. However, the use of high-voltage direct current transmission requires the conversion of power generated by offshore wind turbines, and the establishment of converter stations at sea and on land for power transmission[15]. In this way, the cost and maintenance cost of HVDC transmission are greatly increased. Especially, the construction difficulty, cost and maintenance cost of offshore converter station are much higher than that of land converter station.

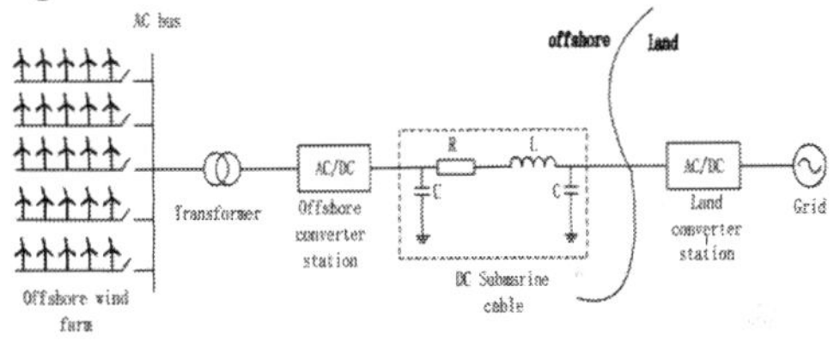

Fig. 2. Flexible DC transmission

Therefore, with the development of China's construction of offshore wind farms with large capacity and long distance, the traditional AC offshore wind power transmission technology will no longer meet the transmission demand, while the grid-connected HVDC transmission system has low practical application cost because of its high cost and complex operation and maintenance.

3 Technical analysis of low frequency offshore energy delivery

In view of the shortcomings of two traditional offshore wind power transmission technologies, low-frequency transmission technology is proposed for large-capacity offshore wind power transmission in the middle and distant sea areas.

3.1 Interconnection of low-frequency offshore power transmission

The structure of a typical offshore low frequency

transmission system is shown in Figure 3. Low-frequency power is generated by offshore wind turbines, boosted by offshore transformers and transmitted to land AC switching stations via submarine cables, then converted to power frequency by frequency conversion devices and integrated into power grid[16-20]. Compared to HVDC system, it can save the platform of offshore converter station, and its construction cost and maintenance cost are greatly reduced[21-25].

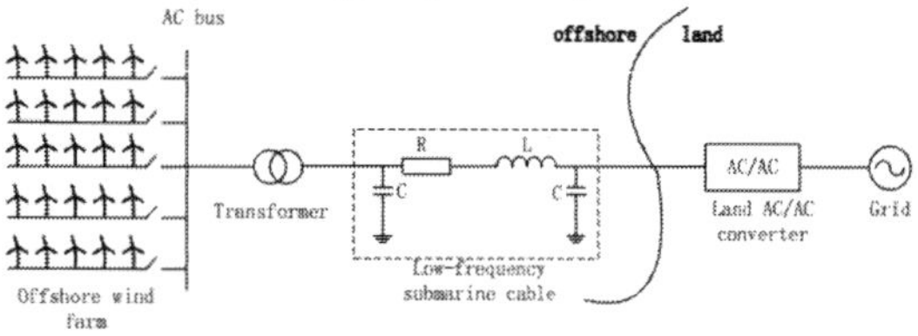

Fig. 3. High voltage low frequency transmission

3.2 Comparison of low frequency and conventional ac delivery mode

By reducing transmission frequency, low frequency transmission system can reduce line reactance and ground capacitance, thus reducing line charging current and shortening electrical distance. Taking 275 kV offshore wind power grid connection as an example, two different types of cables are used to calculate. When the transmission distance reaches 100 km and 175 km respectively at 50Hz, the charging power will occupy the transmission capacity of the whole line and the transmission active power is 0, if the transmission frequency is15Hz, the charging power will be reduced effectively, as shown in Figure 4. Foreign scholars also have similar research, reducing the transmission frequency, the transmission active power decline rate greatly reduced[26].

In addition, the low-frequency transmission system can effectively reduce the terminal voltage fluctuation. Taking the 10kV voltage level and 150 mm2 submarine cable as an example, before the voltage fluctuation reaches the limit, the maximum active power transmission under the low-frequency (15Hz) condition is 1.43 times of the power transmission limit.

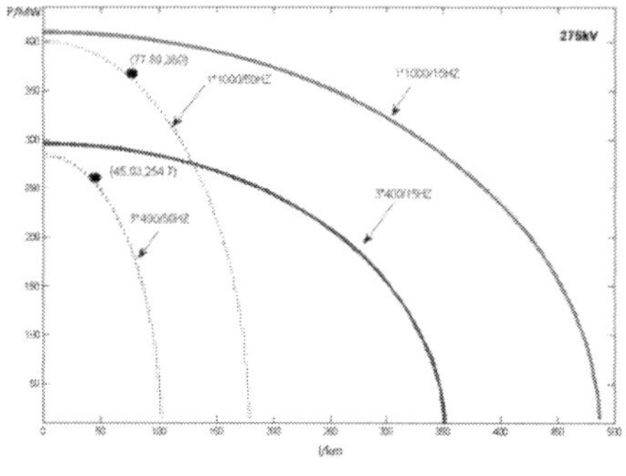

Fig. 4. 275kV Cables transmission power to length

3.3 Comparison of low frequency and flexible dc delivery mode

Flexible DC transmission mode does not have the restriction of transmission distance, but the cost is high. In this paper, two different cases of wind power convergence in deep water are considered. When low frequency transmission or flexible HVDC transmission are used to realize the interconnection of offshore wind power, their primary investment budgets are calculated and compared with each other economically. Generally speaking, there is no offshore converter in low frequency mode, and the cost of offshore converter valve and platform is omitted. However, the low frequency mode is still AC three-phase transmission. Compared with the straight poles, the cost of cable is higher than that of DC. Therefore, low-frequency transmission will have economic advantages when the distance is less than a certain economic distance. Especially with the increase of water depth, the economic advantages of low frequency AC transmission are more obvious.

Following will be a specific example of low-frequency and flexible economy of comparative analysis.

According to ENSTO-E's Report on Offshore Transmission Technology[27], the average installed capacity of the current European offshore wind farms is 400-500 MW. The voltage level of flexible HVDC transmission system is ±320 kV (corresponding AC convergence voltage level is 275 kV). In order to facilitate comparison, the voltage level of low frequency transmission system is 275 kV, and the influence of charging power can be neglected when the transmission frequency is 15 Hz.

Under the scenario of 400 MW wind power outgoing and water depth of 30 meters, the funds for both are detailed in Table 2.

Table 2 400MW/30 meters depth funding comparison (All costs are given in 100 millions of Yuan)

Modes	Offshore AC Platform	Offshore DC Platform	Total Cable Cost/km	Convert er	Transformer	Equivalent Distance
HVDC	5.44	10.28	0.0476	6	3.2	**296.5km**
Low-frequency	8.16 (1)	0	0.07984	2.4	4.8 (2)	

Note (1) Based on the cost of power frequency AC platform, the cost of low frequency AC platform is estimated by volume amplification factor (1.5 times).

Note (2) Based on the cost of power frequency AC transformer, the cost of AC transformer is estimated by the magnification factor of volume and weight (1.5 times).

Therefore, the equivalent distance between low-frequency AC and flexible DC in this case is 296.5 kilometers, when it's less than the distance, Low-frequency AC has more economic advantages.

Water depth is an important factor affecting the cost of offshore platforms. The change of water depth conditions will also change the economic distance between the two modes. Under the scenario of 400 MW wind power outgoing and 50 meters water depth, the funds for both are detailed in Table 3.

Table 3 400MW/50 meters depth funding comparison (All costs are given in 100 millions of Yuan)

Modes	Offshore AC Platform	Offshore DC Platform	Total Cable Cost/km	Conver	Transformer	Equivalent Distance
HVDC	6.3232	12.322	0.0476	6	3.2	**347.3km**
Low-frequency	9.4848 (1)	0	0.07984	2.4	4.8 (2)	

Note (1) Based on the cost of power frequency AC platform, the cost of low frequency AC platform is estimated by volume amplification factor (1.5 times).

Note (2) Based on the cost of power frequency AC transformer, the cost of AC transformer is estimated by the magnification factor of volume and weight (1.5 times).

Therefore, the equivalent distance between low-frequency AC and flexible DC in this case is 347.3 kilometers, when it's less than the distance, Low-frequency AC has more economic advantages.

To sum up, the grid-connected transmission mode of offshore wind farms can be selected by three ways: conventional AC, low-frequency AC and flexible DC. Its technical feasibility distance and economic distance are affected by the capacity, water depth, voltage level, cable cross-section area and model of different wind farms. The optimal transmission mode should be selected according to the specific parameters of wind farms. Generally speaking, when the wind farm capacity is small and the offshore distance is relatively close (generally rated capacity is less than 200 MW, offshore distance is less than 100 km[28]), it is suitable to adopt the conventional AC transmission mode. The transmission distance is short, which can ensure that the voltage fluctuation meets the standard. At the same time, the charging reactive power on the transmission line is within the required range, and there is no need to build offshore or land converter stations, which has low construction cost and maintenance. When the wind farm capacity is large and the offshore distance is less than 100 km to 250 km, it is suitable to adopt low frequency AC transmission mode. On the premise of ensuring voltage fluctuation and charging reactive power to meet the requirements, only one end of land converter station needs to be constructed, which greatly reduces the cost of offshore platform and converter station compared with flexible DC transmission mode, and consequently reduces the maintenance cost and has significant economic advantages. If the offshore wind farm is far away from the shore and exceeds the economic distance of the low frequency transmission mode, the flexible DC transmission mode is suitable.

4 Technical applicability analysis of low frequency offshore wind power delivery technology

In order to verify the technical feasibility of offshore wind power transmission through low frequency, a simulation model of 400 MW fan grid-connected system based on M3C converter valve was built on PASCAD simulation platform.

The system parameters are shown in Table 4, in which the Y side is power frequency network and the X side is 15 Hz low frequency network. The system can realize low frequency wind power Grid-connected function.

Table 4 System parameters

Equipment	Parameters	Figure
M3C converter valve	Rated capacity/*MW*	400
	Rated current/*kA*	0.75
	Submodule Capac-itance Voltage/*kV*	2.4
	Submodule Capac-itance/*mF*	5
	Number of Submodules in Converter Chain	190
	Carrier Frequency/HZ	600
	Current limiting reactance 1/H	0.038
	Current limiting reactance 2/H	0.1
Y-side AC system	Frequency/HZ	50
	Phase voltage/*kV*	120
	Phase current/*kA*	1.125
X-side AC system	Frequency/HZ	15
	Phase voltage/*kV*	120
	Phase current/*kA*	1.125

The system control algorithm structure includes: 1) outer loop active/reactive power control algorithm; 2) inner loop current decoupling control algorithm; 3) outer loop global voltage control algorithm; 4) link modulation wave generation algorithm (including DC voltage control and link DC control); 5) PWM signal generation algorithm.

The system control algorithm can be realized, 1) maintaining the constant active/reactive power of transmission; 2) bidirectional flow of converter energy; 3) decoupling control of active/reactive current; 4) decoupling control of frequency x side of Y side; 5) active power balance of Y side/x side; 6) ensuring the stability of DC voltage of each link section at rated value.

The simulation results are as shown in figure 5.

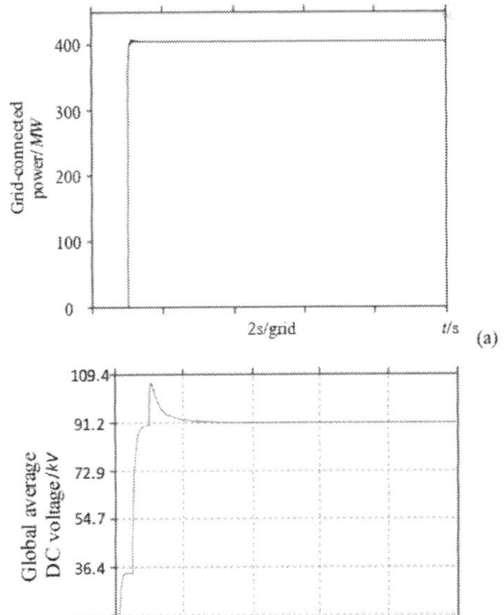

 https://doi.org/10.1051/e3sconf/202018201001

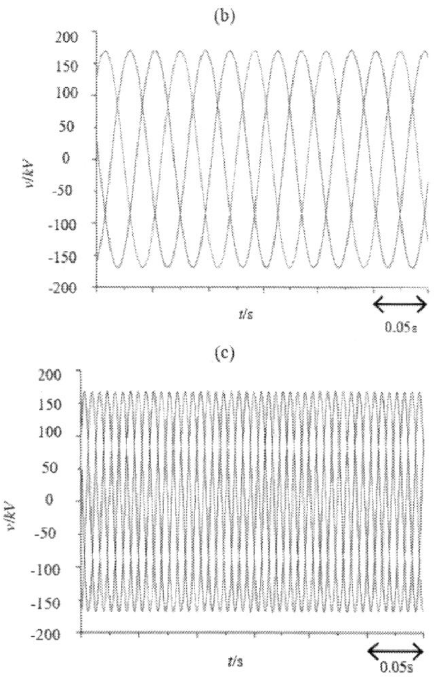

(d)

Fig. 5. Steady-state simulation waveform of offshore wind power Grid-connected system: (a) Grid-connected power and (b) Global average DC voltage and (c) Low-frequency voltage and (d) Power-frequency voltage.

5 Opportunities and challenges of low frequency offshore wind power transmission

Using low-frequency transmission mode to realize large-scale and long-distance offshore wind power transmission can greatly improve the available capacity of submarine cables, and in a certain range of distance, compared with flexible DC transmission mode, the economic advantages are outstanding. From other application scenarios, due to the characteristics of low-frequency transmission technology, it is conducive to the interconnection of multi-terminal network in the islands, and other scenarios such as offshore platform power supply, power supply in remote areas, capacity expansion and transformation of existing lines, improving transmission capacity of lines, increasing power supply radius, avoiding new lines, saving investment and construction. The transmission and power supply problems in the above typical scenarios bring opportunities for the development of low-frequency transmission, a new type of transmission technology.

Through the above analysis, although the low-frequency transmission technology has good technical economy, at present, there are still many technical breakthroughs to be made in practical application.

(1) Low frequency modification of fan side. The offshore wind power low-frequency delivery system described in this paper needs to be reformed at the wind turbine side to achieve low-frequency power generation. The specific implementation method can reduce the gear box speed ratio, simplify the structure of wind turbine, change the operating conditions of wind turbine units, and improve efficiency. Although this method can further reduce the cost of fans, the specific implementation still needs to fully communicate with the fans manufacturers.

(2) Large capacity AC/AC converter. At present, frequency conversion technology is often used in motor drive, but there is no actual development and application of high voltage and large capacity AC converter. Moreover, the number of bridge arms of AC converter based on matrix transformation technology is large and the control process is complex. It is necessary to further study the main electrical parameters and control and protection strategies in combination with application scenarios, device capacity, frequency characteristics and system operation mode.

(3) Primary equipment for low frequency use. At present, the conventional primary equipment is designed according to the power frequency condition. When the frequency decreases, the diameter of the transformer core will increase. At the same time, the flux density of the transformer core will increase with the frequency decreases, leading to saturation of the core easily, distortion of the output waveform, increasing the volume of the core and reducing the flux density of the core are effective ways to avoid this phenomenon. Therefore, in low frequency environment, transformers are required to increase the volume of the core and reduce the flux density of the core. Insulation and structure design need to be re-evaluated; in low frequency environment, the switching time of circuit breaker will be significantly longer than that of power frequency, and the arc-extinguishing ability in this process needs to be further studied; finally, under low frequency fault conditions, the ability of lightning arrester to release energy and valve plate needs to be re-evaluated.

6 Conclusion

(1) The development of offshore wind energy will become an important part of global energy development, and will develop towards large capacity and far sea area.

(2) Considering China's energy structure and the distribution of offshore wind energy resources, we should vigorously promote the development of wind energy in the middle and long distance sea areas of Zhejiang and Jiangsu.

(3) The combination of low-frequency transmission technology and offshore wind power transmission has better technical economy in a certain distance. It can be a beneficial supplement to conventional power frequency AC and flexible HVDC transmission technology.

(4) Low-frequency transmission technology needs further research and promotion. On the one hand, it needs to carry out frequency optimization and networking scheme research in combination with specific scenarios. On the other hand, it needs to cooperate with equipment manufacturers to further study the impact of frequency reduction on primary equipment.

References

1. Global Wind Energy Council. Global Wind Report 2017. Belgium: Global Wind Energy Council, 2018:15-27.
2. Global Wind Energy Council. Global Wind Report 2016. Belgium: Global Wind Energy Council, 2017:1-2.
3. International Renewable Energy Agency. Innovation Outlook: Offshore Wind. Abu Dhabi: International Renewable Energy Agency, 2016:52-95.
4. Liu Yong，Kong Xiangwei，Bai Ke. A Perspective Study On Technical Support System For Building Large Scale Offshore Wind Farm. Resources Science,2009; **31(11)**: 1862-1869.
5. Tan Zhizhong, Liu Deyou, Wang Feng, Yang Zhi. Planning Of Offshore Wind Farms. Shanghai Electric Power, 2007; **2**:125-128.
6. Liu Qi, Xu Yi-Qing. Some Problems On Offshore Wind Power Development In China. Shanghai Electric Power, 2007; **2**:144-148.
7. Fd.Bjx.Com.Cn. Development Status Of Global Offshore Wind Power In 2015 [Online]. Available: Http://News.Bjx.Com.Cn/Html/20160704/747739-2.Shtml. 2016.
8. National Energy Administration. A Reply To The Planning Of Offshore Wind Power In Fujian Province. China: National Energy Administration, 2017:1-2.
9. Guangdong Development And Reform Commission. Development Plan Of Offshore Wind Power In Guangdong Province(2017-2030). China: Guangdong Development And Reform Commission. 2018:7-17.
10. G. B. Power Transmission Over Long Distances With Cables. Presented At: 2004 Cigre.
11. Ning Lianhui, Wang Xifan, Teng Yufei. Experiment On Wind Power Grid Integration Via Fractional Frequency Transmission System. Proceedings Of The Csee, 2011; **31(21)**: 9-16.
12. Chang Yong, Xu Zheng, Zheng Yuping. A Comparison Of The Integration Types Of Large Wind Farm. Automation Of Electric Power System, 2007; **31(14)**:70-75.
13. Jiang Hao, Sun Yu, Liu Hui. Literature Review Of Wind Power Integrated Into Power Systems With Hvdc-Flexible. North China Electric Power, 2017; **1**: 1-7.
14. Wei Xiaoguang, Tang Guangfu, Wei Xiaoyun, Chi Yongning. Study Of Vsc-Hvdc Controller To Mitigate Voltage Fluctuation Caused By Wind Farm Integration. Transactions Of China Electrotechnical Society, 2007; **22(4)** :150-156.
15. Wu Jun-Hong, Ai Qian. Research On Multiterminal Vsc-Hvdc System For Wind-Farms. Power System Technology, 2009; **33(4)**: 22-27.
16. Wang Xifan, Wang Xiuli, Teng Yufei. Fractional Frequency Transmission System And Its Applications. Proceedings Of The Csee, 2012; **32(13)**: 1-6.
17. Chi Fangde, Wang Xifan, Wang Xiuli. New Wind Power Grid Connection Method Based On Fractional Frequency Transmission System. Automation Of Electric Power Systems, 2008; **32(4)**: 59-63.
18. Cho Y, Cokkinides G J, Meliopoulos A P. Lfac-Transmission Systems For Remote Wind Farms Using a Three-Phase，Six-Pulse Cycloconverter. In: Ieee Power Electronics And Machines In Wind Applications(Pemwa), IEEE，2012: 1-7.
19. Teng Yufei, Ning Lianhui, Li Gan. Electromechanical Transient Caculation Method And Transient Characteristics Analysis Of Power System With Fractional Frequency Wind Power System. Automation of Electric Power System, 2015; **39(2)**:67-73.
20. Zhanf Kaoshe, Wang Jianhua, Yao Lixiao. Study On Steady-State Performances Of Power System Containing Fractional Frequency Transmission Lines. Power System Technology, 2001; **25(5)**:10-14.
21. P. B. Wyllie Y T L R. Low Frequency Ac Transmission - Elements Of a Design For Wind Farm Connection. Presented At: 2015 11th Iet International Conference On Ac And Dc Power Transmission.
22. Wang Xifan. Fractional Frequency Power Transmission System. Electric Power, 1995; **(1)**:2-6.
23. Wang Xifan, Wang Xiuli. Study Of Digital Simulation for Fractional Frequency Transmission System. Electric Power, 1995; **(3)**:8-12.
24. Wang Xiuli, Wang Xifan, Wang Jianhua. The Initial Experiment Of Fractional Frequency Transmission System. Electric Power, 1996; **(6)**:33-36.
25. Hu Chaofan, Wang Xifan, Cao Chengjun. Feasibility Research Of Flexible Fractional Frequency Transmission System. High Voltage Engineering, 2002; **3**:16-18.
26. Achara, P. And T. Ise. Operating Phase And Frequency Selection Of Low Frequency Ac Transmission System Using Cycloconverters. Presented At: 2014 Power Electronics Conference (Ipec-Hiroshima 2014 - Ecce-Asia).
27. European Network Of Transmission System Operators For Electricity. Offshore Transmission Technology Report. Belgium:European Network Of Transmission System Operators For Electricity. 2011:35-44.
28. Chen Xia. Wind Power Integration Using Multi-Terminal Hvdc Technology. Wuhan: Huazhong University Of Science And Technology, 2012; 5-7.

Medium and long-term wind energy forecasting method considering multi-scale periodic pattern

Yisha Lin[1,*], *Zongxiang* Lu[1], *Ying* Qiao[1], *Mingjie* Li[2] and *Zhifeng* Liang[2]

[1] State Key Lab of Control and Simulation of Power Systems and Generation Equipment (Department of Electrical Engineering, Tsinghua University), Haidian District, Beijing 100084, China
[2] State Grid Corporation of China, Xicheng District, Beijing 100031, China

Abstract. Medium and long-term weather sequence forecast becomes unreliable beyond two weeks since the weather is a chaotic system. Using values of same months for electricity prediction of wind power is the usual method. This approach defaults wind power output with annual cycle law. However, the periodic pattern can be very complicated in fact with multiple time scales. This paper proposes an approach with multi-scale periodic pattern considered. The application of parametric estimation on cumulative distribution function avoids the difficulty of predicting the power curve. Meteorological condition is considered to some extent via multi-scale periodic pattern explored basing on historical energy data. This work is an exploration for medium and long-term wind power forecasting that can well adapt to existing conditions. It has better prediction accuracy than the method without multi-scale periodicity considered.

1 Introduction

Reliable forecast on mid and long-term generation of wind power can be significant for energy balance, generation scheduling and maintenance scheduling [1] with the penetration of renewable power raising.

However, medium and long-term power timing curves are difficult to obtain under current technical conditions. For one thing, Numerical models for simulation and prediction of atmospheric flows are subject to deterministic chaos and are likely to give unrealistic solutions [2]. The theoretical upper limit of deterministic forecast for weather sequence is about two weeks. After reaching the upper limit of two weeks, the error of deterministic weather forecast is almost equal to the level of natural variability [3]. For another, statistical methods based on historical sequence extrapolation are poorly adaptable [4].

Above problems considered, the main form of wind power forecasting at present for medium and long-term scale is to predict the amount of electricity monthly or yearly. Probabilistic prediction based on principal component analysis and quantile regression is carried out in [5]. However, the prediction quality is strongly correlated with the accuracy of the weather forecast for the next 30 days and so it can be ineffective for areas that are lack of weather forecast meeting the requirements. Gray model is often used in situations over which historical data is insufficient [6], but it cannot handle situation with new installed capacity. Another common practice is to find the representative wind farm of targeted area [7], and then extrapolate the output of the representative wind farm. Forecast on representative wind farm is used as the result for the region. This method is not universally applicable for places with complex terrain.

This paper proposes a forecasting method for monthly electricity generation of wind power. The approach is based on cumulative distribution function (CDF), which avoids the difficulty of predicting the power curve. Multi-scale periodic pattern of wind resource is taken into account, and so the meteorological condition is considered to some extent.

2 Wavelet analysis for multi-scale periodicity

Wavelet analysis is a powerful tool that can reveal the dominant modes of variability. One important feature that makes it distinguished is it decomposes a time series into time-frequency space, and so it can offer both time-domain and frequency-domain information. Because of that, wavelet analysis has better performance in analysing nonstationary time series compared with other signal representation approaches, such as Fourier transform [8]. Wind power series is typically non-stationary with different frequencies at various scales [9], and thus it is ideal to be analysed by Wavelet transform.

Continuous wavelet transform is applied in the following evaluation in view of its good performance on feature extracting [10]. A function that satisfies the admissible condition can be the mother wavelet as the base of transformation. Admissible condition is given as formula (1).

[*] Corresponding author: lys_ncepu@126.com

$$\int_{-\infty}^{\infty} |\omega|^{-1} |\psi(t)|^2 d\omega < \infty \tag{1}$$

The continuous wavelet transform of a signal $f(t) \in L^2(R)$ is defined as the convolution of $f(t)$ with the daughter wavelets $\psi_{a,b}(t)$ which are the translated and scaled versions of the mother wavelet $\psi(t)$.

$$(W_\psi f)(a,b) = \int_{-\infty}^{\infty} f(t)\overline{\psi_{a,b}(t)}dt \tag{2}$$

$$\psi_{a,b}(t) = |a|^{-1/2} \psi((t-b)/a) \quad a,b \in R, a \neq 0 \tag{3}$$

Here, a is scale factor, and b is shift factor. $(W_\psi f)(a,b)$ is also named wavelet coefficient, which indicates the correlation between the wavelet and the input array. Overline indicates complex conjugate.

In this paper, Morlet complex wavelet is chosen to be the mother wavelet. The real part of the wavelet coefficient is convenient for us to study the scale variability over the entire time domain [11].

The wavelet variance Var(a) is the integral of the squared norm of the wavelet coefficients in the time domain, which indicates the periodic wave energy distribution under the different scales. When the scale is fixed, the greater the wavelet variance is, the more obvious the periodic characteristics are, which means that the time series signal has the main period. The wavelet variance is as shown in equation (4).

$$\text{Var}(a) = \int_{-\infty}^{\infty} |(W_\psi f)(a,b)|^2 db \tag{4}$$

3 Electricity generation forecasting method based on cumulative distribution function

Restricted by the accuracy of weather forecasting for long term, we gave up the common idea to calculate generation by the integral of power. We rather choose the CDF as the intermediary, and make the computation concise and easy to process. The transformation from CDF to generation can be expressed as equation (5)

$$W = \tau \int_0^{P_M} (1 - F(x))dx \tag{5}$$

Here, $F(x)$ represents CDF, and P_M is the upper bound of power output, that is capacity in this work. τ represents the time sampling interval.

Since quantities of wind farms were constructed in the recent years, and the old ones are often lack of good data storage conditions. This leads to the situation at present that parametric estimation is chosen to be the method in this paper. Since complex non-parametric estimation is hard to be achieved, which is likely to have better precision. The scheme for forecast of wind power generation is proposed as displayed in Fig.1.

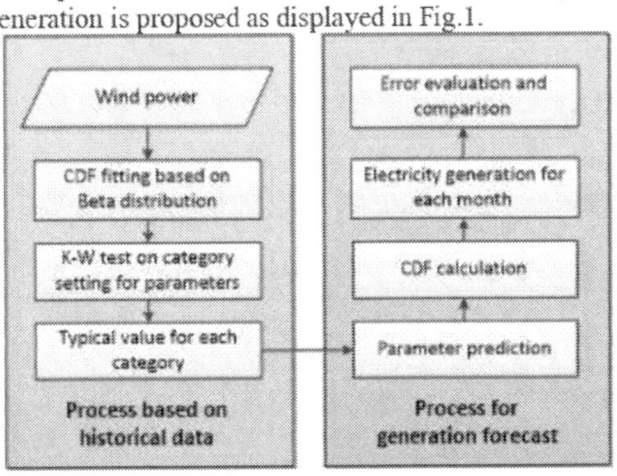

Fig. 1. Overall steps for wind electricity generation.

3.1 CDF fitting based on Beta distribution

Beta distribution is applied for our analysis as it leads to lower error for CDF fitting in most circumstances compared to Weibull distribution and Lognormal distribution, both of which are commonly used [12-14]. This will be reflected in section 4.2.

Probability density function (PDF) of beta distribution is

$$f(x;\alpha,\beta) = x^{\alpha-1}(1-x)^{\beta-1}B(\alpha,\beta)^{-1} \tag{6}$$

$$B(\alpha,\beta) = \int_0^1 t^{\alpha-1}(1-t)^{\beta-1}dt \tag{7}$$

Correspondingly, CDF of beta distribution is

$$F(x;\alpha,\beta) = \int_0^x t^{\alpha-1}(1-t)^{\beta-1}dtB(\alpha,\beta)^{-1} \tag{8}$$

Maximum Likelihood Estimation is used to obtain the value of two parameters of Beta distribution: α and β.

3.2 Parameter prediction based on periodicity

Parameter estimation is expected to be done according to the multi-scale periodic pattern from existing data. That is, several significantly different categories will be distinguished based on periodicity, and then typical values will then be figured out for predicting.

Assume that there are m major periods found, and then at least $C_m^1 + C_m^2 + ... + C_m^m$ classification schemes are optional. Kruskal-Wallis nonparametric test (K-W test) is used to find the best division, for which the differences between categories are most significant. If null hypothesis that distribution of different category is the same cannot be rejected, it means that the class definition method is not reasonable.

Once categories are settled, the median of parameters for each class is defined as the typical value. If no effective classification based on periodicity is obtained, then other information, for example, the values of previous months will be adopted directly.

3.3 Evaluation indices

The difference of empirical CDF obtained from historical data and estimated CDF reflects the error of forecast. Root mean square error (RMSE) is used to quantify the deviation.

$$RMSE = \left(\frac{1}{N}\sum_{i=1}^{N}(F(x_i;\alpha,\beta) - \hat{F}(x_i))^2\right)^{1/2} \tag{9}$$

Here, $\hat{F}(x)$ refers to empirical CDF.

Relative error (*RE*) is applied to measure the offset of forecasting results for electricity generation.

$$RE = (W - \hat{W})\hat{W}^{-1} \tag{10}$$

Therein, $\hat{W}$ represents the actual electricity generation.

4 Case Study

Case study is designed based on the data of wind farms in north Hebei province. Time ranges from 2015 to 2017, and the time step is 1 minute.

4.1 Multi-scale periodic pattern

Since our goal is to analyze the medium and long-term pattern of wind power, which corresponds to monthly or even yearly scales, we choose series of monthly wind power generation as input. Electricity generation also reflects the abundance of wind resources. Contour map over the real part of wavelet coefficients is shown in Fig.2.

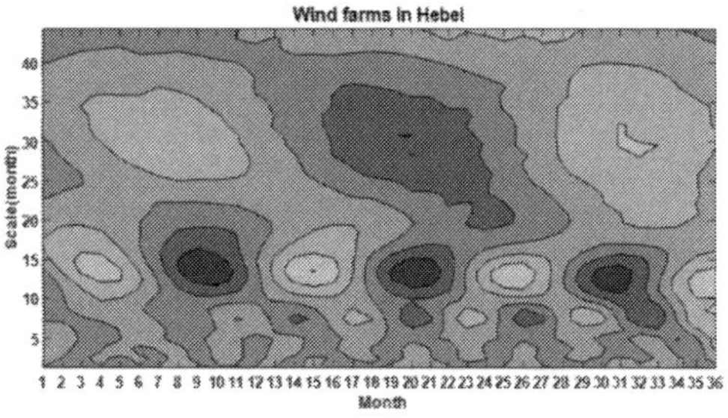

Fig. 2. Contour map based on wind power generation of Hebei.

In the figure, the horizontal axis represents time in the unit of month and the vertical axis represents the wavelet scale. If the region of the contour map is filled with the warm color, the real part is positive which means that the wind resource is abundant, and the wind power generation is vast. Opposite conclusion will be made if the filling color is cold.

The wavelet variance of wind farms in Hebei is exhibited in Fig.3 to figure out the main periodicity. The leading periodic component is 13.54 months, the second-domain component is 29.54 months, and the third is 7.385 months. They are reflectively defined as medium, long and short period for the following discussion.

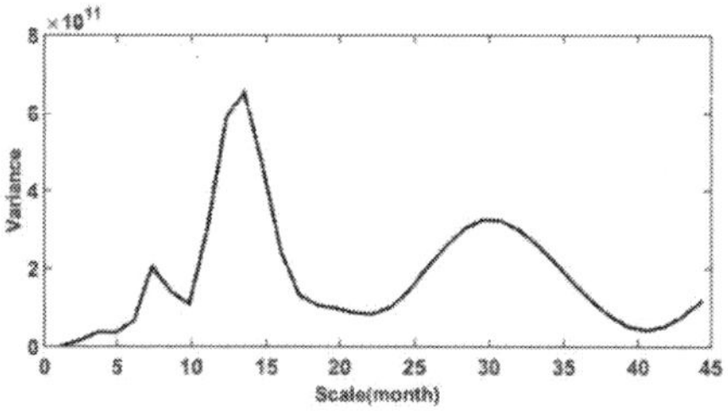

Fig. 3. Wavelet variance diagram.

4.2 CDF fitting

Distributions that are widely used for wind power are tried for each month, including Beta distribution, Weibull distribution and Lognormal distribution. RMSE for each situation is shown in Fig.4, and illustrates that applying Beta distribution has best performance for this case.

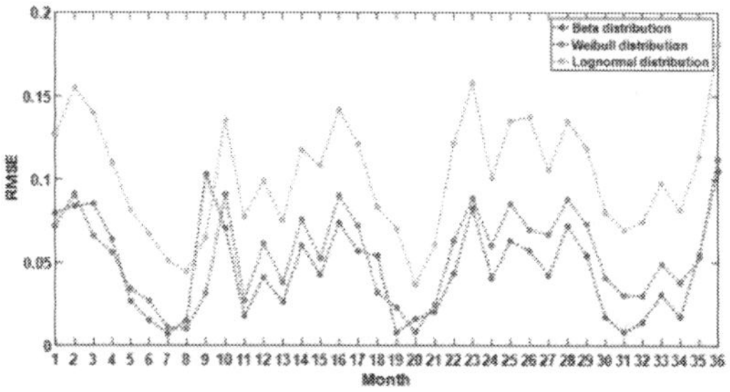

Fig. 4. RMSE for CDF fitting based on different distributions.

Table 1. Parameter estimation and RMSE of CDF fitting for 2015.

Month	1	2	3	4	5	6
α	1.981	0.725	2.915	2.068	1.477	0.859
β	10.672	5.079	13.267	11.101	6.985	5.005
RMSE	0.067	0.071	0.071	0.054	0.027	0.015
Month	7	8	9	10	11	12
α	0.877	0.734	1.002	1.063	0.915	0.968
β	7.582	7.478	3.194	3.356	6.736	3.992
RMSE	0.004	0.009	0.083	0.069	0.018	0.043

Estimation of parameters for Beta distribution and RMSE for CDF fitting are given as Table 1(results for 2015) and Table 2(results for 2016). The maximus RMSE is 0.085. It indicates that the method to forecast electricity generation using CDF based on Beta distribution is feasible.

Table 2. Parameter estimation and RMSE of CDF fitting for 2016.

Month	1	2	3	4	5	6
α	1.609	1.368	1.819	1.452	1.676	0.617
β	6.056	4.481	5.455	3.880	4.443	3.095
RMSE	0.029	0.059	0.046	0.074	0.059	0.053
Month	7	8	9	10	11	12
α	0.883	0.705	0.772	1.289	2.467	1.284
β	5.806	8.941	4.762	3.833	4.696	3.946
RMSE	0.013	0.010	0.015	0.051	0.085	0.045

4.3 Parameter prediction

K-W test is applied to different modes, and statistical significances are listed in table 3 for parameter α and β. Significance level is set as 0.05. Long, medium and short period was defined in section 4.1.

For every period considered, there are two classes, one is that the wind resource is abundant, and another is not. For example, if "Long-Medium" periods are concerned, there are four categories for parameters. The first is that wind resource presents abundant for both periods, and the second shows scarce characteristic for both periods. Other two correspond to the situations that wind resource is ample for one period, and necessitous for another period.

Table 3. Significance for different classifications.

Periods considered for classification	Statistical significance for α	Statistical significance for β
Long	0.835	0.728
Medium	0.002	0.321
Short	0.08	0.448
Long-Medium	0.01	0.638
Medium-Short	0.006	0.196
Long-Short	0.354	0.084
Long-Medium-Short	0.048	0.116

Whether takes medium period as the base for categories affects the conclusion a lot. When analyzing on the Medium-Short mode, we found situation that wind resources are in shortage for both periods (defined as category1 in Fig.5) has significantly different data distribution with others (defined as category2 in Fig.5). We then made two categories according to the result, and the statistical significance became 0.001, lower than any as shown in Table 3. It is what we use in the following analysis. The distribution of α for different group is displayed in Fig.5.

α for category 1 is obviously small than category 2. We select medians as the typical value for each class, which is 0.7527 and 1.4101 separately.

As we can refer from Table 3, difference is not significant between different categories no matter the way of defining groups. This indicates that the parameter prediction method according to periodic pattern is not suitable for β. Hence, we use the average of the last two months to be the predicted value for β as the lack of quantitative data.

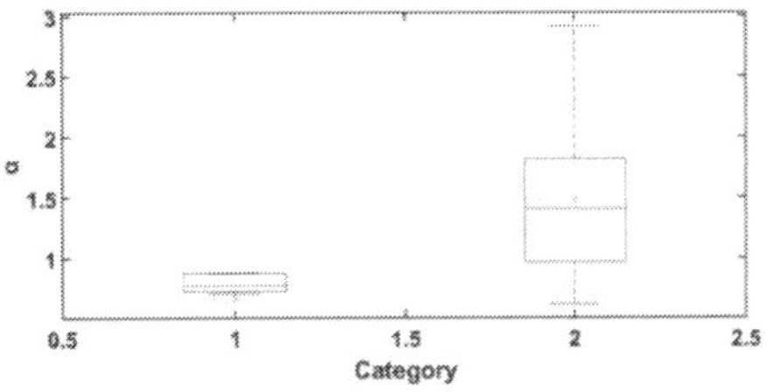

Fig. 5. Distribution of for different category.

4.4 Wind electricity generation forecast

Since the category belonging is very important for selection of α, forecast on abundance situation for medium and short periods should be done firstly. Fig.6 shows the abundance condition for months of 2015 and 2016, and it is consistent with Fig.2. The red square corresponds to the warm part in Fig.2, and the blue one to the cool tone. The duration of stages of abundant and dry wind resources shows a strong regularity. Therefore, we do random sampling to forecast the mode for the future months according to the distribution reflected in the historical data. Predicted results are displayed as hollow squares.

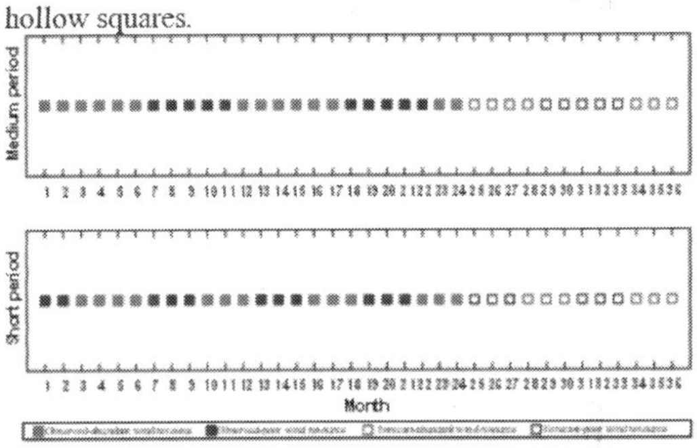

Fig. 6. Forecast on abundance of wind resources.

Parameters can be set then for months that generation is expected to be predicted. To reflect the preponderance with multi-scale periodic pattern concerned for forecast, we make the benchmark that parameters are set only based on the data of same months in the previous years. The absolute values of RE for electricity generation forecast for 2017 is demonstrated in Fig.7.

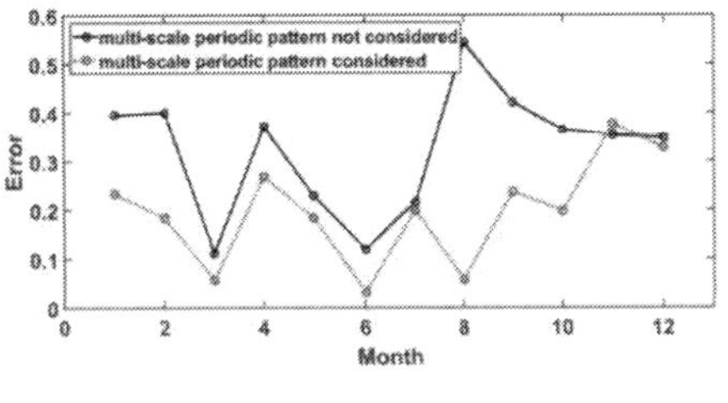

Fig. 7. Forecast error of electricity generation.

The mean error with multi-scale periodic pattern not considered is 0.3226, and the value drops to 0.1958 with the method proposed in this paper. However, the error is higher than the benchmark in November, the probable cause is that power dispatching institution increase the accommodation of wind power to fulfill the promise of reducing wind abandon made in the early of the year.

Error of electricity generation forecast and CDF fitting is compared as Fig.8 shows. The larger the error of CDF fitting, the higher the absolute value of RE for electricity generation. Moreover, the prediction error of electricity generation amplifies the error of CDF fitting during integral operation.

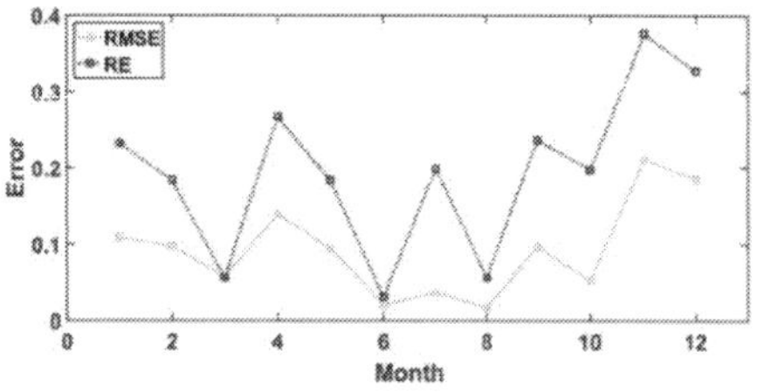

Fig. 8. Error of electricity generation forecast and CDF fitting.

5 Discussion and conclusion

This work provides a forecast method for monthly electricity generation of wind power. The method is put forward under the condition that no credible weather sequence forecast for medium and long term could be referred. Multi-scale periodic pattern is in consideration to predict the parameters of Beta distribution, which is chosen to fit the probability density distribution of wind power. The method has better prediction accuracy than the method without periodic pattern considered, since the mean relative error of electricity generation decrease from 0.3226 to 0.1958.

However, this paper is just a preliminary work for medium and long-term wind power forecasting. The absolute value of relative error reaches 0.3901 for the worst situation, and it is not a satisfying result for practice. Further efforts could be made for the following points:

- Use complicated distribution, for example, mixture distribution, with accumulated data increasing.
- More information is deserved to be explored from historical data, not only periodicity for wind resources.
- Try to integrate weather forecasting information to make prediction for wind electricity generation.

The research is supported by Science and technology project of STATE GRID Corporation of China: research on statistical rules of wind power generation characteristics and its application (SGNXDK00DWJS1800013).

References

1. S. Han, Y. Qiao, J. Yan, Y. Liu, L. Li, Z.Wang. Appl. Energy. **239**,181-91 (2019).
2. A.Mary Selvam, S. Fadnavis. Chaos Solitons & Fractals. **10**,1577-82 (1999).
3. F. Zhang, Y. Sun, L. Magnusson, R. Buizza, S. Lin, J. Chen, et al. Journal of the Atmospheric Sciences. **76**, 1077-91 (2019).
4. D. Alencar, C.Affonso, R.Oliveira, J. Moya Rodríguez, J. Leite, J. Reston Filho. Energies. **10**, 1976 (2017).
5. D. Li, Z.Ren, W. Yan, J.Zhu, X. Zhao, J.Yu. Zhongguo Dianji Gongcheng Xuebao. **37**, 5238-47 (2017).
6. H. Citakoglu, A. Aydemir. 2019 IEEE Jordan International Joint Conference on Electrical Engineering and Information Technology (JEEIT); 2019 9-11 April 2019.
7. Z. Guo, Y.Dong, J.Wang, H.Lu. Math. Probl. Eng. **2010**, 17 (2010).
8. P. Chowdhury, B.N. Dwivedi. Planet. Space Sci., **67**, 92-100 (2012).
9. X. Bao, C. Li. Journal of Wind Engineering and Industrial Aerodynamics, **193**, 103982 (2019).
10. A.C. Furon, C. Wagner-Riddle, C.R. Smith, J.S. Warland. Agricultural and Forest Meteorology, **148**, 1305-17 (2008).
11. X. Chen, L. Yin, Y. Fan, L.Song, T.Ji, Y. Liu, et al. Sci. Total Environ., 134244 (2019).
12. X. Yuan, C. Chen, M. Jiang, Y. Yuan. Applied Soft Computing, **82**, 105550 (2019).
13. P. Wais. Renewable Energy, **103**, 15-29 (2017).
14. I. Pobočíková, Z. Sedliačková, M. Michalková. Procedia Eng, **192**, 713-8 (2017).

Mid-term Scenario Generation for Wind Power Using GAN with Temporal-correlation Enhancement Block

Yupu Ma[1,2,*], *Ming* Ma[3], *Ningbo* Wang[3], *and Ying* Qiao[1]

[1]State Key Lab of Control and Simulation of Power Systems and Generation Equipment (Dept. of Electrical Engineering, Tsinghua University), Haidian District, Beijing 100084, China
[2]School of Electrical Engineering, Northeast Electric Power University, Jilin 132012, Jilin Province, China
[3]Wind Power Technology Center of Gansu Electric Power Company, Lanzhou Province 730070, China

Abstract. Under the background of increasing renewable energy penetration and rigid demand for mid-term generation planning, the accurate expression of wind power uncertainty becomes more and more important. Firstly, this paper analyses the problem of scenario generation models based on traditional Generative Adversarial Networks(GAN), point that the fluctuation of scenarios that it generated usually deviates greatly from the real one. And further proposes a convolutional structure, that called Temporal-correlation Enhancement block (TE block), which can solve the aforementioned problem by enhance the temporal correlation perception ability of convolutional layers. Then, the problems of traditional conditional Wasserstein GAN-Gradient Penalty(WGAN-GP) in mid-scale scenario generation are discussed, and a conditional WGAN-GP scenario generation model that is suitable for mid-term scenario generation is presented. This model-free non-parametric model can generate a large number of realistic wind scenarios efficiently according to the given conditions. In order to validation, we use the data collected in a province in northern China to train the model in the Case Study part, and compare it with the traditional model. Result shows the fluctuation distribution deviation problem is improved obviously on the model with TE block, and in the comparison of auto-correlation coefficient, the proposed model also outperform the traditional model. This verifies the superiority of the proposed model in temporal expression ability compared with the traditional model, as well as the feasibility of the mid-term wind scenario generation.

1 Introduction

In recent years, the rapid development of installed capacity of renewable energy in China has brought great challenges to the accommodation of renewable energy. Wind power, as one of the most uncertain power sources, is more prominent. On the one hand, in China's new energy center, 72% of the units are large-scale thermal power units over 300 MW[1]. The minimum start-up and shutdown time of these units is long, and the start-up and shutdown cost is high, so the overall flexibility is insufficient. Therefore, it is necessary to formulate the start-stop plan of these units in the mid-term. On the other hand, in order to avoid the uncertainties brought by wind power, the guaranteed capacity of wind power is usually calculated according to the year-round statistics. [2] points out that when the guaranteed rate is 95%, the guaranteed capacity coefficient can even be reduced to less than 0.4%, which leads to wind power being almost ignored on the start-stop and maintenance plan, which will affect the wind accommodation seriously. Therefore, in order to solve the above contradictions, the uncertain modelling of wind power in the mid-term scale is becoming more and more important.

Scenario method is a commonly used method to express the uncertainty of wind power. To be more specific, it is to generate a series of wind scenarios according to the historical data distribution of wind power. Each scenario represents a possible hypothetical reality to realize the simulation of physical uncertainty. The generated scenarios can be used in different optimization models and prediction models. In [3], a new scenario generation method is proposed to deal with the uncertainty of wind power forecasting, and then the wind scenario is used into the stochastic security-constrained unit commitment (SCUC) model. In [4], neural networks are used to generate wind scenarios. Each possible scenario generated in this way contains wind power ramping information, and the distribution of specified Wind Power Ramp Event (WPRE) attributes can be obtained randomly according to the scenarios.

Current scenario generation methods mainly include parametric and model-based methods, such as [5], [6], and nonparametric method. These parametric methods are based on the assumption that wind power obeys a certain probability distribution, and it can express the temporal-spatial correlation of wind power to a certain extent. However, it is difficult to reproduce the high-order and high-nonlinear characteristics of wind power

* Corresponding author: 293596479@qq.com

by using these low-order models, which leads to inaccurate modeling of the spatial-temporal characteristics of real wind power distribution. Nonparametric methods do not require wind power to obey a distribution assumption. For example, in [7], radial basis function neural networks (RBF) algorithm is used to transform the generated scenarios into a series of quantiles to represent the uncertainty of wind power.

Recently, with the emergence of Deep Learning technology, more and more researches in the field of scenario generation has begun to focus on a powerful generative model, Generative Adversarial Networks (GAN)[8]. This is a novel unsupervised learning algorithm based on nonparametric method. The algorithm does not directly learn some explicit distribution, but generates synthetic data close to the real distribution of train dataset through the game of two neural networks. In [9] , for the first time, GAN with Wasserstein distance is introduced to generate renewable energy scenarios, which is a data-driven, model-free and non-parametric scenario generation method. It proves that GAN has high feasibility in expressing temporal-spatial correlation and uncertainty of wind power. In [10], the conditional WGAN-GP algorithm is introduced for the first time. The prediction error category is predicted by SVM with the wind power prediction curves, and the scenario with this kind of prediction error is generated.

The above scenario generation based on GAN algorithm can accurately simulate the distribution of wind power, but the convolutional layer is used as the feature extraction method in their models. This leads to the convolution-based GAN algorithm cannot accurately express the probability distribution of fluctuations in sequence generation. In this paper, we first introduce the Temporal-correlation Enhancement block (TE block) as a supplement to the inherent defects of convolutional layer. At the same time, the above scenario generation studies all limit the time scale to the short-term time scale which is predominant in the past, so this paper then analyses the problems that conditional WGAN-GP model may encounter in the mid-term scale, and gives a set of solutions. Finally, a series of case studies are given to demonstrate that the proposed TE block model improves the expression ability of temporal correlation in generating scenarios, and the specific effect of mid-term wind scenario generation is shown at last.

2 Temporal-correlation enhancement block

2.1 Problems in traditional GAN scenario generation structures

Previous scenario generation based on GAN, including WGAN model and WGAN-GP model, whose discriminators are all based on convolutional neural network. Convolutional network structure is more suitable for feature extraction with translational invariance. However, wind power curve is a kind of time series. The operation of moving summation in convolutional layers will destroy the correlation between adjacent elements. From the result of the final generated curve, it can be seen that the generated curve and the original curve are in good agreement with the probability distribution of the wind power, but the fluctuation distribution of the correlation which reacts sequential correlation will deviate greatly. This inherent defect of convolutional networks will make the generated curves more fluctuating than the actual ones, which will have a negative impact on some occasions where the uncertainty of fluctuation needs to be accurately modeled. And when it comes to the Mid-term scale, this phenomenon will be more obvious with the increase of time scale and the fluctuation rate.

2.2 Explanation of the TE block

Since the moving summation of the traditional convolutional layer will destroy the temporal correlation, the temporal correlation is artificially reconstructed by the Temporal-correlation Enhancement (TE) block using different time scale differences in this section.

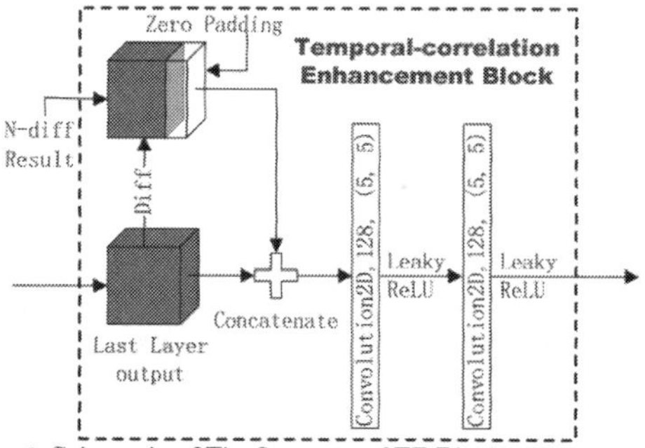

Fig. 1. Schematic of The Structure of TE Block.

The schematic of the proposed TE block is shown in Fig. 1. The TE block still adopts the convolutional structure, so its input is also the feature maps $F_{w \times h \times c}$ of the whole batch output of the last layer.

Firstly, the TE block reshapes the feature maps F to the series S as shown in (1).

$$F_{w \times h \times c} \rightarrow S_{m \times c} \tag{1}$$

Where w, h, c and m are the width, height, number of channels of the F and the length of the reshaped series S, respectively.

Then 1-th to N-th order difference (N-diff) of reshaped S is performed N times as shown in (2).

$$s^N_{m_0,c_0} = s_{m_0+N,c_0} - s_{m_0,c_0}, m_0 \leq m - N \tag{2}$$

where, s is a sample in the S, and is used to form the fluctuation series s^N on the time scale of N. After N-diff operation, the length of the these series is form $m-N$ to m, respectively, so it is necessary to do the zero-padding to let the short series be as long as the original one.

Finally these N-diff series is reshaped to $w \times h \times c$ and concatenated (or summated if the amount of channels is too large), by the channel dimension to form new feature map tensors, which will act as the input of the convolutional layer later. So the TE block could output the convolutional feature maps reflecting the difference feature directly.

3 Mid-term wind scenario generation based on conditional WGAN-GP

3.1 Conditional WGAN-GP

Conditional WGAN-GP algorithm is a kind of generative model. Its training process can be regarded as a game between two neural networks. When the game reaches Nash equilibrium, one of the neural networks can transform noise into the conditional synthetic data. The two neural networks are connected sequentially and named generator and discriminator respectively. The training methods are as follows:

First, sampling produces noise $z \sim p(z)$, where $p(z)$ denotes the distribution of noises, which usually obey Gauss distribution. The difference of conditional GAN is that samples need to be labelled, a batch of samples (x, y) are taken out, the labels of samples y, usually one-hot classed, are linked together with z. The generated data $x_g = G(z \mid y, \theta^G)$ can be obtained by the generator, where $G(\cdot, \theta^G)$ represents the generator with current parameters. Then take x, x_g and y into the discriminator to get the discriminator outputs $V_T = D(x \mid y, \theta^D)$ and $V_F = D(x_g \mid y, \theta^D)$.

1) Discriminator Training: freeze θ^G while training θ^D, use Wasserstein-l as the loss function, and add an gradient penalty term additionally. The discriminator's loss function is:

$$\begin{aligned} L_D = & \mathop{\mathbb{E}}_{x_g \sim \mathbb{P}_g}[D(x_g \mid y, \theta^D)] \\ & - \mathop{\mathbb{E}}_{x \sim \mathbb{P}_r}[D(x \mid y, \theta^D)] \\ & + \lambda \mathop{\mathbb{E}}_{x_{av} \sim \mathbb{P}_{av}}\left[\left(\left\|\nabla_{x_{av}} D(x_{av})\right\|_2 - 1\right)^2\right] \end{aligned} \tag{3}$$

Where x_{av} is the result of random weighted averaging of x and x_g. The loss function of WGAN-GP not only trains the discriminator to distinguish whether the two distributions are consistent, but also discriminates the size of their difference. The Wasserstein-l loss function can improve the gradient disappearance or explosion problems in traditional GAN training. [11] mentions that Wasserstein-l is founded on the premise that the neural network satisfies the 1-Lipschitz constraint, while the literature [12] introduces the gradient penalty (GP) term, which is the third term in (3),to meet that requirement.

2) Generator Training: When the discriminator is trained n_{critic} times, freeze θ^D training θ^G, similarly use y, z to generate V_F. Backpropagation based on the generator's loss function, enabling the generator to generate synthetic data that is sufficient to fool the discriminator. The specific loss function of the generator is as follows:

$$L_D = -\mathop{\mathbb{E}}_{x_g \sim \mathbb{P}_g}[D(x_g \mid y, \theta^D)] \tag{4}$$

Overall, if we ignore the GP term, the training process is as follows:

$$\min_G \max_{D \in \mathcal{D}} \mathop{\mathbb{E}}_{x \sim \mathbb{P}_r}[D(x)] - \mathop{\mathbb{E}}_{x_g \sim \mathbb{P}_g}[D(x_g)] \tag{5}$$

3.2 Mid-term wind scenario generation

3.2.1 The drawback of traditional conditions GAN in mid-term scale

In the past, [10] has used conditional GAN for one-day scenario generation with the day-ahead wind power prediction or clustering results as its one-hot conditions. This method is very suitable for single-day conditional scenario generation, but when the time scale comes to the mid-term, directly using this method may bring a series of problems:

First of all, the forecast error of wind power in the mid-term is larger than day-ahead, and even in some cases there is no point predictions for wind power, so the mid-scale scenario generation should use the energy forecast of wind power as the basis.

More importantly, the number of one-hot encoded classes will increase exponentially as time scale increases. Assuming that there are c categories for single-day wind power, n-day wind power requires at least c^n classes to have the same performance as single-day scenario generation. Because of this problem, when the time scale is relatively long, the Classification condition method will greatly reduce the feasibility of the algorithm.

In addition, if clustering result is used as conditions, the conditions will not have an obvious physical meaning, and cannot be continuously changed as needed, resulting in inflexibility in practical applications.

3.2.2 The model overview

For the above reasons, it is appropriate to use the interval of daily/total energy of wind power as the condition. The model overview is shown in Figure 2. The model uses the TE block to replace the convolutional layers in generator and discriminator respectively to ensure the fluctuation of different time scales could be reflected. The rest parts of model are consistent with the conditional WGAN-GP.

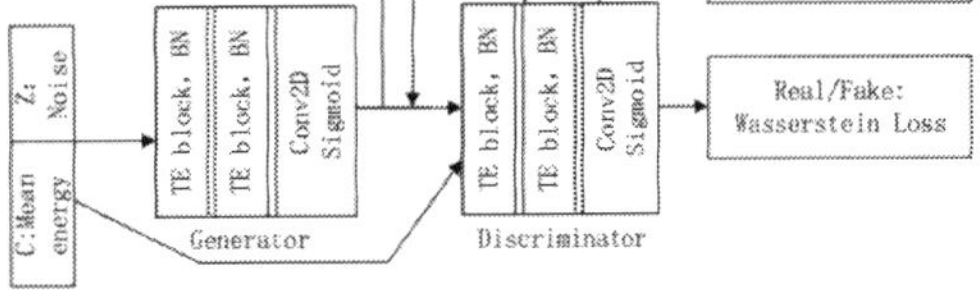

Fig. 2. The overview of Mid-term Scenario Generation Model.

4 Case study

Because this paper mainly discusses how to enhance the temporal correlation modeling ability of the generative models, the spatial correlation between multiple wind fields is not considered as the key point. Therefore, the dataset uses the annual total wind power data collected from a province in northern China. The total installed capacity is 7300MW. Moreover, the wind power of a larger spatial scale is more representative of the mid-term characteristics of wind power than the short-term. Since the time scale is mid-term, a resolution of 1h is considered sufficient. The length of the train samples in this paper is 5 days (120h), but as a matter of fact the proposed model structure can be applied to various non-extreme time scales.

4.1 Model details

The training samples are obtained for each successive 120h on the annual wind power curve and slide one hour every time to form a sample matrix, which is then normalized to enable the neural network to converge correctly. The model input is the concatenation of the condition C and noise Z , and the dimension of noise is also set to 120 here. The mode of the input condition is the interval category of the total energy of wind power per day/total time resolution. And it is repeated and extended to the same length as the noise to make it easier to concatenate. As shown in Fig. 3 (a) classified by daily time and (b) classified by total time. The wind energy interval is evenly divided according to the quantile of the daily/total-time energy of wind power. For the sake of brevity, this paper only divides the energy into three parts, and the threshold is (1970MW/3405MW) × Total time, respectively.

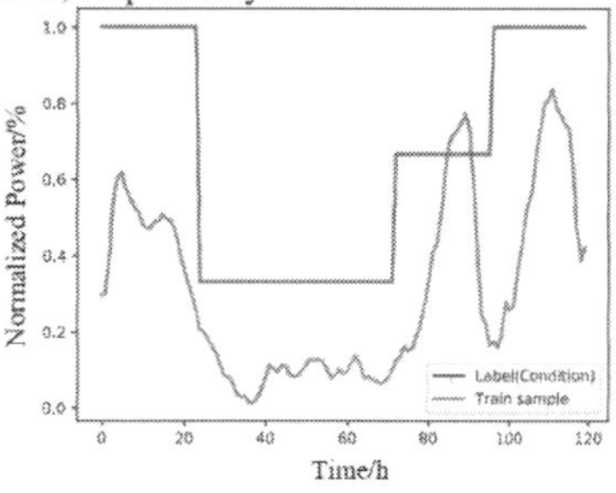

(a)

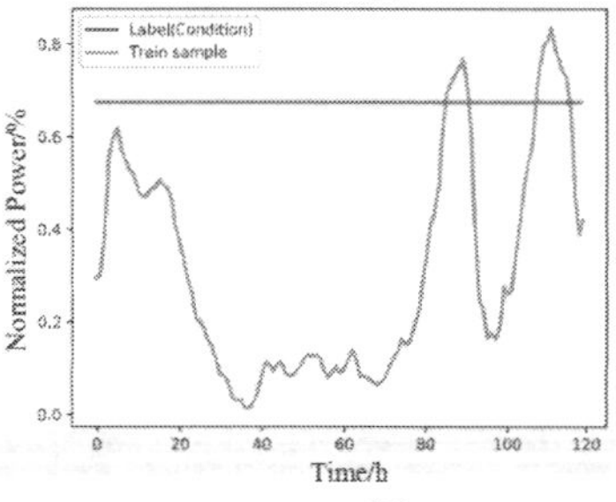

(b)

Fig. 3. The daily/total-time condition formation mode.

4.2 Comparison of learning effects between two models

In order to test whether the proposed model can improve the ability of uncertainty modelling of the generated scenarios' fluctuation, and ensure that the specific distribution generating ability of the.generative model does not decrease at the same time. The conditional probability distribution of the power and the fluctuation of the generated scenarios are compared in this section, to tell if the model is able to reproduce the real data distribution. In order to achieve a sufficient amount of data for each category, = total-time wind energy classification is used as its condition.

Figures Fig.4(a) and Fig.4(b) show the probability density distributions of the first and second classes of the original sample probability density distribution, the distribution of generated scenarios output by the WGAN-GP model without/containing the TE block with corresponding conditions. By comparing the three curves, we can see that the two models have better learning effects and can accurately reflect the distribution of wind power under different conditions.

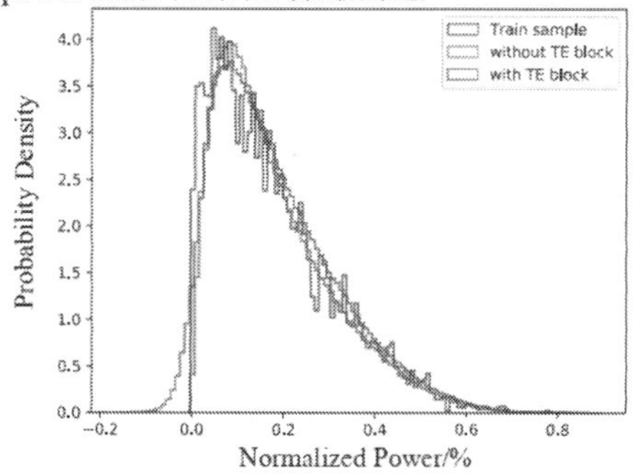

(a)

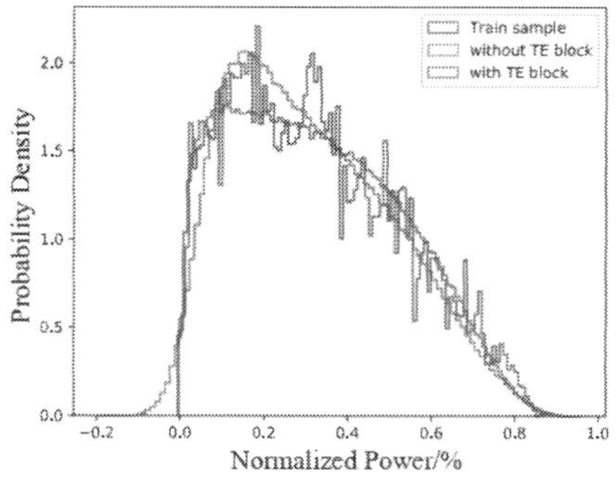

(b)

Fig. 4. Learning results of models on different classes of wind power distribution.

Fig. 5(a) and Fig. 5 (b) respectively show the conditional probability density distribution of the first and second class of original sample wind power fluctuations, and the fluctuation distribution of scenarios generated by the WGAN-GP model without/containing the TE block. It can be clearly seen that the green curve (denotes the model containing the TE block) is more close to the real distribution, and the wind power fluctuation distribution of the original WGAN-GP generated scenarios deviates significantly from the real distribution. From the point of view of fluctuation distribution, it is proved that the proposed model is more capable of capturing temporal correlation.

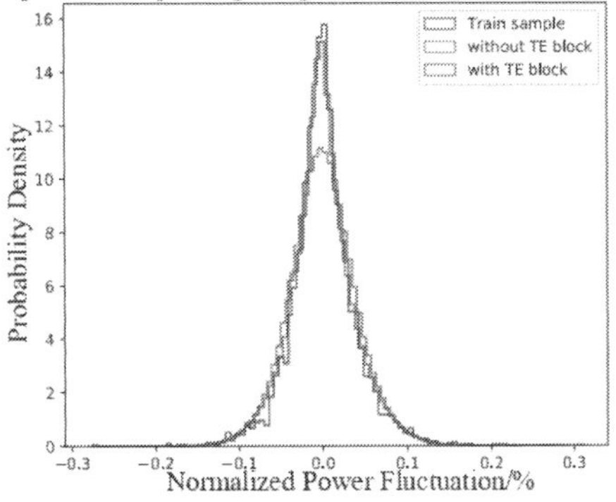

(a)

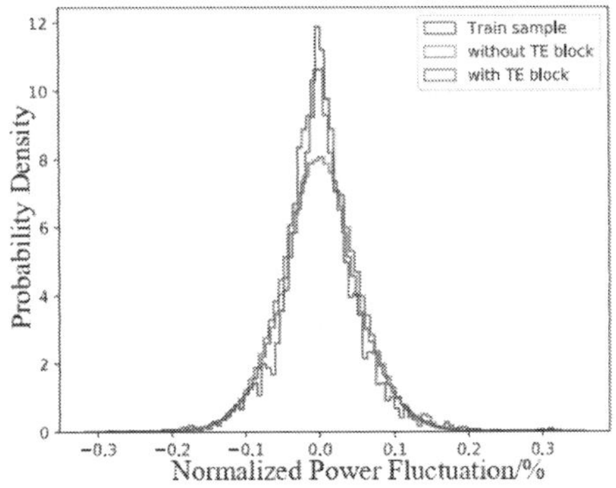

(b)

Fig. 5. Learning results of models on different classes of fluctuation distribution.

In addition, the auto-correlation coefficients which can represent the temporal correlation to some extent are also counted. By using traversal search to generate similar scenarios nearest to the train samples, the MAE distribution of the autocorrelation function is counted as shown in Fig. 6(a). It can be seen that the ACF function of the model with TE block is closer to the real situation. And Figure 6(b) gives a more specific case study.

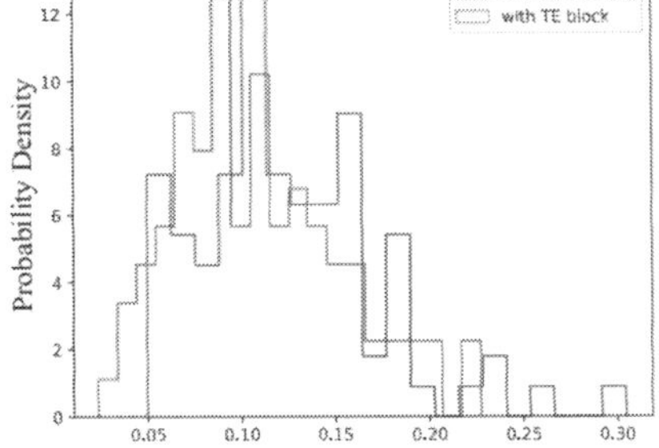

(a)

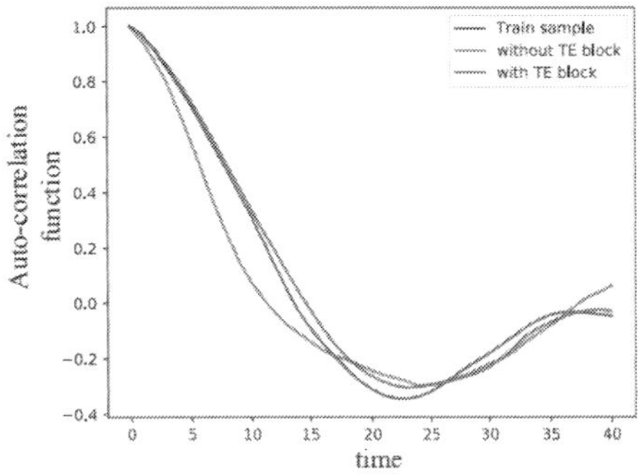

(b)

Fig. 6. Auto-correlation coefficients comparing with its nearest sample.

4.3 Specific scenarios generation

Using the condition of daily classification, scenario generation is performed on two different conditions of 100h and 8000h respectively to simulate the uncertainty of wind power. The training samples corresponding to 100h and 8000h (blue thick lines) are shown in fig. 7 (a) and fig. 7(b), and the input conditions are also shown for the 8000h samples.

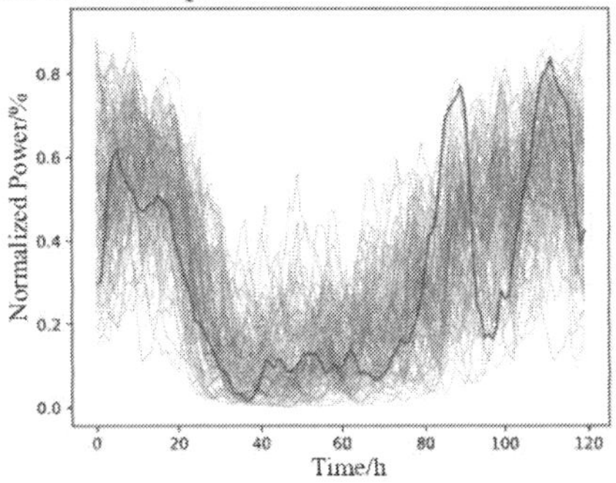

(a)

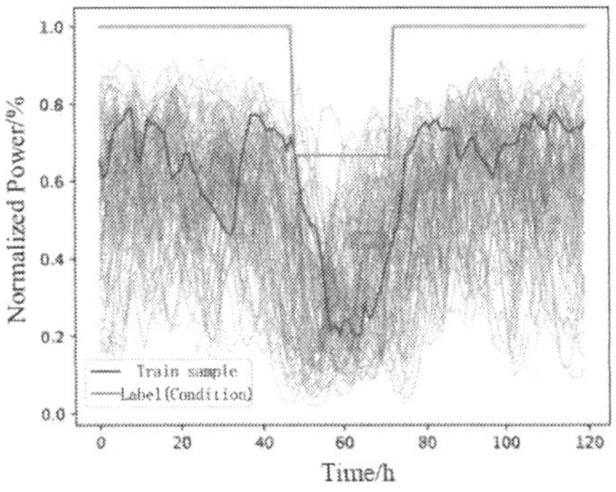

(b)

Fig. 7. A large number of scenarios generating with a given condition.

5 Conclusion

Scenario generation is an important tool for decision-making in the power system with high wind penetration rate. Firstly, this paper discusses the insufficient expression of temporal correlation in the common conditional GAN algorithms, and proposes a solution based on modified neural network structures in the model. Then, the possible problems of applying conditional WGAN-GP algorithm in the mid-term are studied, and a scenario generation model based on improved conditional WGAN-GP model for mid-term wind scenario generation is established. Finally, examples are given in Case Study to verify the feasibility of the proposed model in scenario generation. Compared with the model without TE block, the proposed method improves the learning ability in temporal correlation such as fluctuation, and a series of generated mid-term wind power scenarios is shown in the end.

This work was supported by the Gansu Youth Science and Technology Fund Project (18JR3RA011); Gansu Provincial Major Project (17ZD2GA010).

References

1. State Grid Corporation. Gansu Power Grid 2018 Annual Operation Mode [R].2018.
2. Huang Honglin, Song Lili, Zhou Rongwei, et al. Characteristics Analysis of Wind Power Fluctuations for Large-scale Wind Farms [J]. Proceedings of the CSEE, 2017(06):26-37.
3. Quan H , Srinivasan D , Khosravi A . Incorporating Wind Power Forecast Uncertainties Into Stochastic Unit Commitment Using Neural Network-Based Prediction Intervals[J]. IEEE Transactions on Neural Networks and Learning Systems, 2014, 26(9):2123-2135.
4. Cui M , Ke D , Sun Y , et al. Wind Power Ramp Event Forecasting Using a Stochastic Scenario Generation Method[J]. IEEE Transactions on Sustainable Energy, 2015, 6(2):422-433.
5. Ma, Xi-Yuan, Yuan-Zhang Sun, and Hua-Liang Fang. "Scenario generation of wind power based on statistical uncertainty and variability." IEEE Transactions on Sustainable Energy 4.4 (2013): 894-904.
6. Morales J M, Mínguez R, Conejo A J. A methodology to generate statistically dependent wind speed scenarios[J]. Applied Energy, 2010, 87(3):843-855.
7. Sideratos G , Hatziargyriou N D . Probabilistic Wind Power Forecasting Using Radial Basis Function Neural Networks[J]. IEEE Transactions on Power Systems, 2012, 27(4):1788-1796.
8. Goodfellow I J , Pouget-Abadie J , Mirza M , et al. Generative Adversarial Networks[J]. Advances in Neural Information Processing Systems, 2014, 3:2672-2680.
9. Chen Y , Wang Y , Kirschen D S , et al. Model-Free Renewable Scenario Generation Using Generative Adversarial Networks[J]. IEEE Transactions on Power Systems, 2018:1-1.
10. Zhang, Yufan, et al. "Typical wind power scenario generation for multiple wind farms using conditional improved Wasserstein generative adversarial network." International Journal of Electrical Power & Energy Systems 114 (2020): 105388.
11. Arjovsky M, Chintala S, Bottou L. Wasserstein GAN[J]. 2017.
12. Gulrajani I, Ahmed F, Arjovsky M, et al. Improved Training of Wasserstein GANs[J]. 2017.

LSTM-based Short-term Electrical Load Forecasting and Anomaly Correction

Lei Zhang[1,*], *Linghui* Yang[2] , *Chengyu* Gu[1], *Da* Li[3]

[1] State Grid Shanghai Electric Power Research Institute, Department of Power Source, Shanghai 200437, China
[2] State Grid Shanghai Electric Power Research Institute, General Manager Office, Shanghai 200437, China
[3] State Grid Shanghai Electric Power Research Institute, Department of Human Resource, Shanghai 200437, China

Abstract. The Emergence of the Ubiquitous Power Internet of Things(UPIoT) facilitates data sharing and service expansion for the power system. Based on the architecture of the UPIoT and combined with deep learning technology, short-term electrical load forecasting and anomaly correction could be used to improve the overall performance. Since short-term electrical loads are non-linear and non-stationary [1] and could be easily affected by external interference, traditional load forecasting algorithms cannot recognize the correlation between the time sequence thus rendering low prediction accuracy. In this article, a Long Short-Term Memory (LSTM) based algorithm is proposed to improve the prediction accuracy by utilizing the correlation between the hourly load sequence. Then, the real-time forecasting outputs are compared to the raw data in order to detect and dynamically repair the anomaly so as to further improve the performance. Experiment results show that the proposed approach outputs low Mean Square Error (MSE) of around 0.2 and could still hold it at around 0.3 with corrected data when the anomaly is detected, which proves the accuracy and robustness of the algorithm.

1 Introduction

Electrical load forecasting plays an important role in the construction of smart grid system. Since the electric load forecasting relies on a large amount of historic data, the credibility of the time series of original load sequence has a great influence on the prediction results of any prediction method. Performing real-time anomaly correction and load forecasting simultaneously could improve the reliability of the raw data thus improving the accuracy and stability of the prediction algorithm. Therefore, the study of load prediction and anomaly repair algorithm is of great significance to improving the overall performance of the smart grid.

In the 1960s, the research into load forecasting was gradually leaning to the practical application. Load forecasting methods could be mainly categorized into two types: traditional load forecasting technology and intelligent forecasting technology [2]. Traditional power load prediction algorithms include MA, ES and ARIMA [3]. Since these algorithms are established on the assumption that the observed and the future time series are linearly correlated, they are not effective when processing time series with significant nonlinear characteristics and therefore, have no learning or self-adaptive abilities. In recent years, the load forecasting technology has entered a new era of intelligent prediction and an enormous amount of research regarding machine-learning-based load forecasting has ensued gradually matured. For instance, in [4] the authors use a support vector machines based regression model coupled with empirical mode decomposition to for long-term load forecasting. In [5] electricity demand is forecast using a kernel based multi-task learning methodologies. In [6], the authors used Artificial Neural Network (ANN) ensembles to perform the building level load forecasting. In [7, 8], Recurrent Neural Network (RNN) and Long Short-Term Memory (LSTM) based algorithms are proposed and varied to achieve better overall prediction results. Though enormous number of approaches have been proposed to improve the prediction results, little research has been done to diagnose the reliability of the training data which could influence the forecasting performance to a great extent if an anomaly exists in the historical time series.

This article proposes an approach based on a four layer LSTM model for short-term load forecasting and anomaly correction. LSTM fully considered the correlation of the time series data and perfectly solved the "gradient vanishing" problem of the RNN model. In addition, by comparing the prediction results with the real data, anomalous load data in the historical time series can be detected and corrected before the next prediction iteration. In this sense, the forecasting system could not only provide accurate prediction results, but also warn the power enterprise to detect abnormal conditions of power load in time which significantly increase the practicability of the algorithm. Experiment results show that the mean square error(MSE) between the prediction and the real data is around 0.2 and has a higher prediction

* Corresponding author: lei.zhang8@mail.mcgill.ca

accuracy compared to RNN or other traditional algorithms. In addition, when the anomaly is induced, the MSE between the prediction results using the repaired data and the real data is still around 0.3, which proves the robustness and stability of the proposed approach.

2 An outline of the LSTM

As the simplest type of neural network, Perceptron was developed in the 1950s and 1960s by Frank Rosenblatt [9], inspired by earlier work by Warren McCulloch and Walter Pitts. Perceptron comprises only one input layer and one output layer. Following that, more complex models containing multiple hidden layers have merged. In this work, we mainly focus on the application of the back propagated neural network designed for training time series, i.e., the input of a hidden includes the output of the previous layer at this moment along with its own output at the previous moment. Typical algorithms include RNN, LSTM, etc.

2.1. The RNN architecture

RNN was first proposed by Jeffrey l. Elman in 1990 [10]. Since RNN is more capable of using the information from the past to predict the future, it is widely used in natural language processing, time series processing, voice processing and other time series related fields. A standard RNN architecture is shown in Fig. 1 where the transmission process for RNN is unfolded by time period. $x(t)$ represents the input at time t, $o(t)$ represents the output at time t, and U, V and W denotes the input, output at time t and the weight output at time $t-1$, respectively. Each arrow represents one transformation, which means that the arrows are attached with weights. As shown in the unfolded architecture of RNN, the neurons in between the hidden layer are also connected with weights in standard RNN structure which implies that the status of a hidden layer at a previous moment would affect the status of itself at a later moment as time goes on.

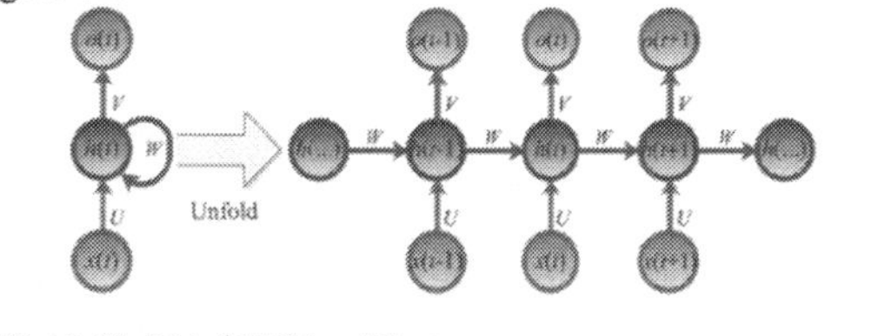

Fig. 1. Unfolded RNN architecture

The data flow of the standard RNN also possesses the characteristics of: 1) weight sharing, i.e., the U, V and W in the figure share the same value. 2) each input only establishes a weighted connection with its own route rather than other neurons. Based on the above characteristics, the output of RNN at time t can be expressed as:

$$o(t) = \text{softmax}(Vh(t) + c) \tag{1}$$

$$h(t) = f\left(Ux(t) + Wh(t-1) + b\right) \tag{2}$$

where $f(\cdot)$ is the activation function of the hidden layers, generally tanh or sigmoid function.

Although RNN is able to learn the data correlation in time, it suffers from the long-term dependency problem, i.e., when the time interval between relevant information and the prediction position grows large, the historical data cannot be traced due to "gradient vanishing" or "gradient exploding", resulting in the loss of learning ability.

2.2 The LSTM architecture

In order to solve the problems of "gradient vanishing" and "gradient exploding" for RNN, Hochreiter & Schmidhuber [11] proposed a variant of RNN named as LSTM in 1997, which could learn long-term information and is widely implemented after improvement and promotion. As shown in Fig. 2, LSTM also contains a chain structure of recurrent modules. In standard RNN, the recurrent module is only composed of a simple structure such as a tanh layer. Alternatively, The LSTM module is composed of three different structures, namely, gated cell where gated means that the cell decides whether or not to store or delete information (e.g. if it opens the gates or not), based on the importance it assigns to the information. The assigning of importance happens through weights, which are also learned by the algorithm. This simply means that it learns over time which information is important and which not. A typical LSTM has three different types of gates: input, forget and output gate. These gates determine whether or not to let new input in (input gate), delete the information because it isn't important (forget gate) or to let it impact the output at the current time step (output gate).

With the evolution of neural networks, more popular variants of LSTM were invented such as peephole connections [12] and Gated Recurrent Unit (GRU) [13], which are widely used in various learning tasks related to time series. **NOTE** that the proposed approach in this article is based on the most basic LSTM structure.

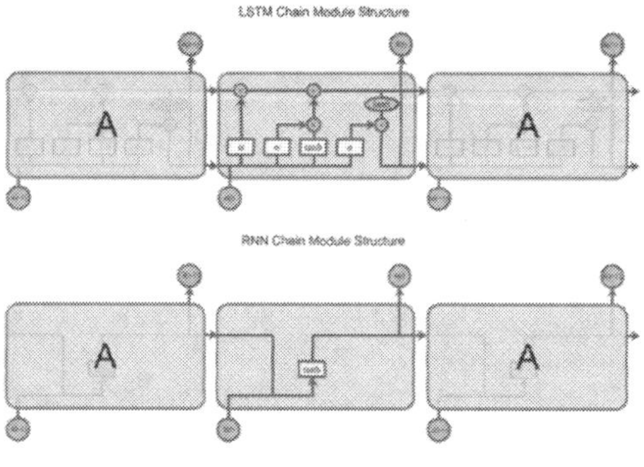

Fig. 2. LSTM and RNN Chain Module Structure

3 Using LSTM for load forecasting and anomaly correction

3.1. Load forecasting

Electrical load forecasting is a kind of prediction based on the historical load data. Normally, before conducting

 https://doi.org/10.1051/e3sconf/202018201004

load forecasting, a large amount of historical data should be collected. The data should be wisely chosen since the more standard and representative the training data, the more accurate the predicted value could be obtained. After that, the original training data need to be normalized and reshaped to accelerate the convergence and the fit the model dimension. Then the model would be trained with appropriate parameters. After the model is trained, the predicted value could be obtained by passing the historical load data to the trained LSTM model. The complete training process of load forecasting using LSTM is shown in Fig. 3.

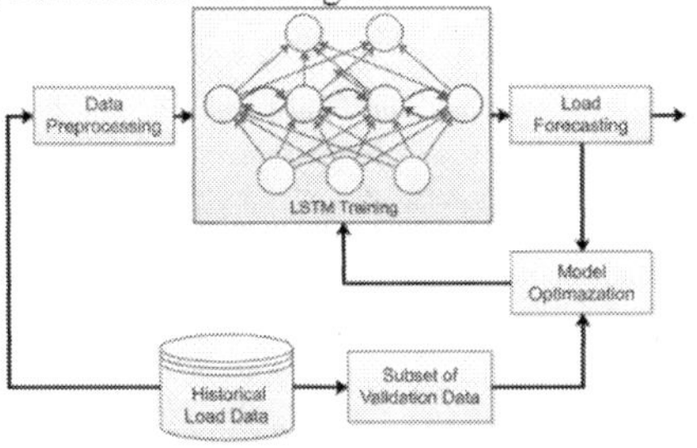

Fig. 3. The training process of load forecasting using LSTM

3.2 Anomaly correction

The anomalous data would affect the prediction results negatively since the pattern of the load sequence is destroyed and the prediction model could no longer recognize the relationship between data. Since LSTM model is distinct for its strong robustness, a small amount of anomaly will not have a serious impact on the predicted results. Therefore, the predicted results can be used to identify anomaly and correct it by the prediction results. Common anomalous data could be categorized into 3 types [14]:

- **Missing data**: Mainly due to the malfunction of the SCADA system, data in the database appears as null values.
- **Deficient data:** Mainly due to the line maintenance, equipment maintenance or electric meter damage in a certain period of time. The load curve increases and decreases drastically in a day or a period of time compared with the normal load of the adjacent day or the planned load of the day.
- **Minimax data:** The load exceeds the peak or valley value in a day dramatically in a non-peak or non-valley moment.

4 Numerical results

Experiments are conducted using Keras library in python.The dataset is the hourly power consumption in Toronto Canada from January to July 2016. After the LSTM model is trained, standard dataset is used for testing, and the accuracy of the prediction results was analyzed in terms of MAPE and MSE. In order to test the anomaly correction results for the proposed approach, corruption of the standard data (missing values, minimax values, etc.) is introduced artificially to perform prediction under anomaly situations. The prediction results are then compared with the real value at the that moment to determine whether the anomaly needs to be fixed. Finally, the prediction results before and after correction are analysed and a conclusion is drawn. The complete diagram of the proposed approach is shown in Fig. 4.

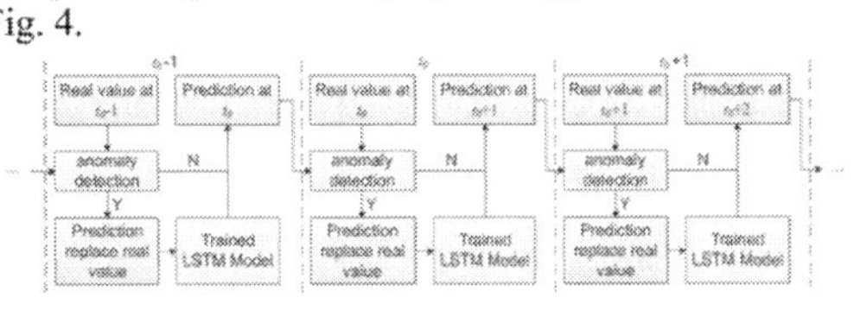

Fig. 4. Complete diagram of the proposed approach

4.1 System setup

Raw data is standardized by subtracting the mean of the original data from all the training data. Since the proposed approach uses the previous 22 load values to predict the 23rd load value, the 22 values before each moment are set to be the training features, and the load value at that moment is set to be the training label.

Since the correction results of the missing data and deficient data are similar, only the missing data case and the minimax case will be discussed in the following context. For the construction of missing data, the load value for a random period of time is assigned to be 0 in the test data. In this experiment, five consecutive random time series in the test dataset are set to be 0 while the others remained unchanged. For the construction of the minimax data, a random time point is selected from the test data. Then the previous moment of this time is set to be a random large number, and the later moment of the time point is set to be a random small number.

4.2 Prediction results for uncorrupted data

Fig. 5 shows the prediction results for uncorrupted test data (487 sets in total). Intuitively, the prediction result is very close to the real data. In order to statistically evaluate the accuracy of the prediction, Mean Absolute Percentage Error (MAPE), Mean Square Error (MSE) and Root Mean Square Error (RMSE) of the prediction and the real data are calculated and compared with RNN prediction results. The results are shown in Table 1.

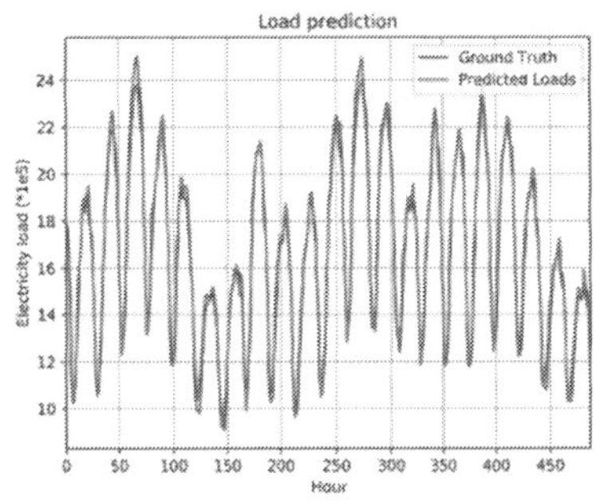

Fig. 5. Prediction results for uncorrupted data

https://doi.org/10.1051/e3sconf/202018201004

Results shows that the overall performance of LSTM model is better than the RNN with both high prediction accuracy and short computational time.

Table 1. Prediction results for LSTM and RNN

Model	MSE	RMSE	MAPE	CPU time
LSTM	0.256	0.506	2.639	0.023s
RNN	0.350	0.591	3.712	0.025s

4.2 Prediction results for corrupted data

4.2.1 The prediction results for missing data

Fig. 6(a) and 6(b) shows the prediction results over the uncorrected anomalous data and the corrected anomalous data, respectively. Obviously, the missing data has greatly affected the accuracy of prediction in period of time when corruption occurred. However the influence of corruption is getting weaker as the time sequence deviated from anomaly point which indicates that the proposed algorithm has fair anti-interference ability. Furthermore, when the dataset is fixed, it can be seen that the difference between the corrected data and the ground truth data is relatively small and the prediction results are more accurate than the results without correction, especially the results in the time period with missing data.

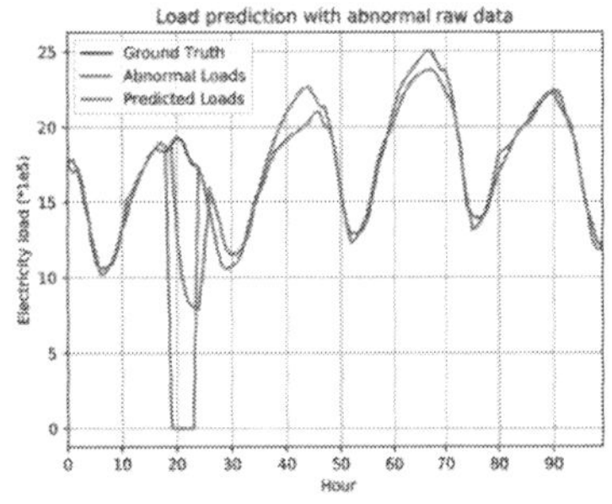

(a). The prediction results over uncorrected data

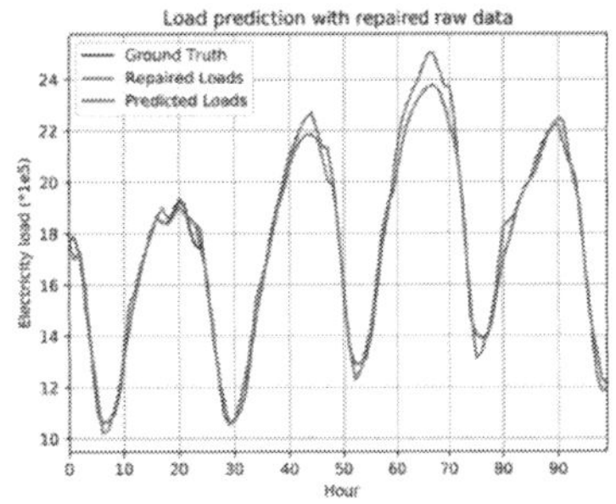

(b). The prediction results over corrected data

Fig. 6. The prediction results for missing data

Numerical results are also presented in Table 2 to further discuss the performance. As shown in Table 2, the proposed anomaly correction algorithm results in smaller MSE and MAPE but longer CPU time due to the dynamic correction during prediction. In addition, although the prediction accuracy over the uncorrected data has decreased, it has not decreased significantly, which also proves the robustness of the system from another perspective, i.e., certain amount of interference does not cause serious damage to the stability of the system.

Table 2. Prediction results for missing data

Data	MSE	RMSE	MAPE	CPU time
Standard	0.256	0.506	2.639	0.023s
Uncorrected	1.214	1.101	3.433	0.019s
Corrected	0.269	0.519	2.678	0.073s

4.2.1 The prediction results for minimax data

Fig. 7(a) and 7(b) shows the prediction results over the uncorrected minimax data and the corrected minimax data, respectively. Similar to the missing data senecio, when the dataset is corrected, the difference between the corrected data and the ground truth data is relatively small and the prediction results are more accurate than the results without correction. However, it can be intuitively seen that the influence of minimax value on prediction is less than that of missing value. This phenomenon could possibly be explained by the sensitivity of the LSTM model to the correlation of continuous time series. Since the missing data is a series of anomaly in continuous time while the minimax data is just some discrete points over time, the damage of the data correlation is more serious for missing data which result in inferior prediction performance.

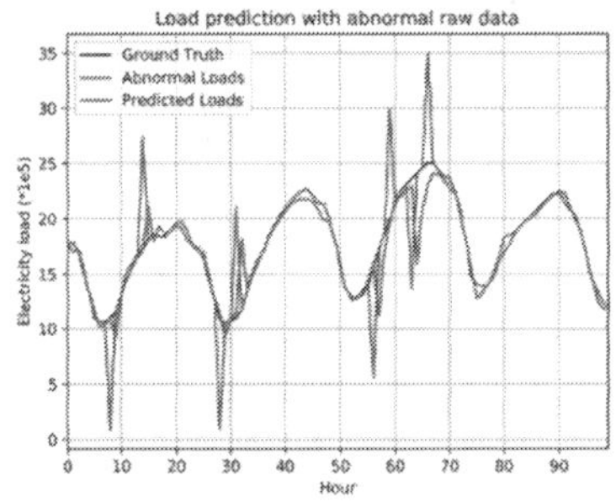

(a). The prediction results over uncorrected data

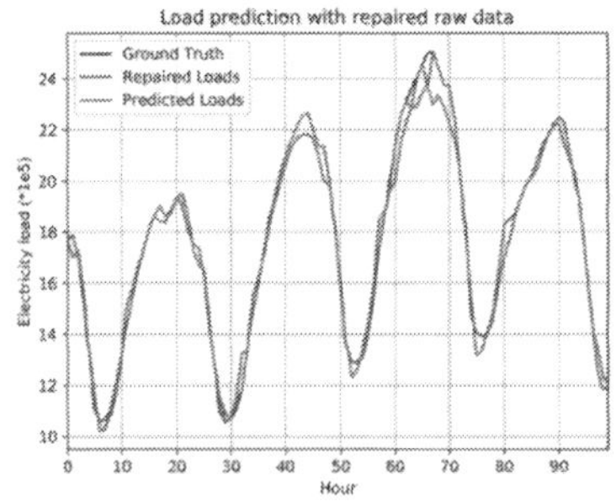

(b). The prediction results over corrected data

Fig. 7. The prediction results for minimax data

Numerical results are also presented in Table 3 to further discuss the performance. As shown in Table 3, the proposed anomaly correction algorithm results in smaller

MSE and MAPE and is even smaller than the results for missing data which further illustrates the sensitivity of the proposed model to data correlation.

Table 3. Prediction results for minimax data

Data	MSE	RMSE	MAPE	CPU time
Standard	0.256	0.506	2.639	0.023s
Uncorrected	0.550	0.741	2.912	0.025s
Corrected	0.271	0.520	2.644	0.054s

5 Conclusions

Short-term electricity load is hard to predict due to its nonlinearity and non-stationarity. However, the traditional non-learning algorithm cannot capture the correlation of load series in the time dimension rendering low prediction accuracy. In this paper, a LSTM-based algorithm is proposed and tested for short-term load prediction and anomaly correction on the basis of a large amount of existing research. The results show that the proposed approach could learn the correlation of load series and output accurate predictions with a short convergence time. In addition, when anomalous data such as missing data and the minimax data appears, the algorithm is able to detect the occurrence of anomaly and repair the anomaly adaptively. The error between the prediction result over corrected data and the standard data is small, which indicates that the proposed model has the advantage of high accuracy, stability and practical significance.

Furthermore, the algorithm of the prediction results over minimax data is better than that of the missing data in general which means that the algorithm has certain sensitivity for continuous anomaly sequence over time and cannot adapt to the data distortion over a long period of time. Therefore future research needs to be carried out into how to better improve the proposed model to make it less sensitive to long time data corruption is worth thinking. Finally, due to the existence of dynamic anomaly correction, the processing time of the algorithm is increased. How to improve the convergence speed remains to be studied.

References

1. Q. W. Ran, Y. Z. Shan, Q. Wang, et al., Wavelet neural network-PARIMA method for short-term load forecasting of power system, *Chinese journal of electrical engineering* **23(3)**, 38-42 (2003)
2. Y. Du, Z. Guo, L. Z. lu, et al., Overview of short-term power system load forecasting methods, *Technology and market* **5**, 339-340 (2015)
3. D. Alberg, M. Last, Short-Term Load Forecasting in Smart Meters with Sliding Window-Based ARIMA Algorithms, (2017)
4. L. Ghelardoni, A. Ghio, D. Anguita, Energy Load Forecasting Using Empirical Mode Decomposition and Support Vector Regression, *IEEE Transactions on Smart Grid* **4(1)**, 549-556 (2013)
5. J. B. Fiot, F. Dinuzzo, Electricity Demand Forecasting by Multi-Task Learning, *IEEE Transactions on Smart Grid*, 1-1 (2016)
6. J. G. Jetcheva, M. Majidpour, W. P. Chen, Neural network model ensembles for building-level electricity load forecasts, Energy and Buildings 84, 214-223 (2014)
7. J. Ku, R. Goomer, A. K. Singh, Long Short Term Memory Recurrent Neural Network (LSTM-RNN) Based Workload Forecasting Model For Cloud Datacenters, Procedia Computer Science 125, 676-682 (2018)
8. Y. H. Zhang, C. M. Qiu, X. He, et al., A Short-Term Load Forecasting Based on LSTM Neural Network[J], Electric Power Information & Communication Technology, (2017)
9. F. Rosenblatt, Perceptron Simulation Experiments, Proceedings of the Ire **48(3)**, 301-309 (1960)
10. J. L. Elman. *Finding structure in time*, Cognitive Science **14(2)**, 179-211 (1990)
11. S. Hochreiter, J. Schmidhuber, *Flat Minima* **9(1)**, 1 (1997).
12. G. Gelly, J. L. Gauvain, Optimization of RNN-Based Speech Activity Detection, *IEEE/ACM Transactions on Audio Speech & Language Processing* **99**, 1 (2017).
13. K. Kang, H. B. Sun, C. K. Zhang, et al., Short-term electrical load forecasting method based on stacked auto-encoding and GRU neural network, *Evolutionary Intelligence*, (2019)
14. Z Y Tian, M L Zhang, R Zhao. Identification and correction of abnormal load data in short-term power load forecasting. *Jilin electric power* **6**, 25-27 (2004

A Study on the Potential of Electricity Generation from Cassava Rhizome in Eastern Region Area of Thailand

Mananya Saengsaior[1,*], *Weerachai* Arjharn[1], *Pansa* Liplab[2], *Thipsupin* Hinsui[2]

[1]School of Agricultural Engineering, Institute of Engineering, in Suranaree University of Technology, Thailand
[2]Center of Excellence in Biomass, Suranaree University of Technology, Thailand

Abstract. In this study, the objective was to investigate the quality and suitability of processing, improving the quality of waste biomass from cassava growing in east region area of Thailand. There is a method for studying quantitative data from the land use database together with the geographic information system. And the physical quality and chemical properties of cassava rhizomes have been studied from the analysis of cassava rhizome samples in the laboratory. The results of the study found that east region area of Thailand is the biggest area for cassava cultivation in Thailand. Crop residual ratio (CRR) of cassava rhizome was 0.112 with the total remaining amount of 636,003.35 tons/year (~53.40% wb moisture content and 8.19 MJ/kg heating value). It was estimated that the total amount of cassava rhizome ready to use as biomass with chip/pellet grade at moisture content of ~10.17% wb. and heating value of ~12.50 MJ/Kg is 89,040 tons/year, having equivalent energy of 1,113.01 GJ or 26.58 ktoe. This could be used to produce electricity of 61.8 GWh or used to supply power plants with total capacity of 8.8 MW.

1 Introduction

Cassava is one of the important agricultural crops in Thailand, and it is among the five agricultural produces' export. According to the Office of Agricultural Economics. Cassava or tapioca is an important agricultural commodity in Thailand. Also, majority of Thai farmers own their farm land spread over rural areas. The reason is that nature of cassava can be harvested all over the year and can be grown in most suitable lands in Thailand. In 2001, Thailand's agricultural trade ranked 10th in the world, and that accounted for 11.7% of agricultural exports [1] Thus, the Thai government policy was to promote exporting agricultural products, which the international trade data of Thailand in 2008. Thailand is the third-largest producer of Cassava in the World [1,2]. In 2018, 24 million tons of processed cassava pellets were produced, with an approximate value of 44 percent of all cassava grown in Thailand.

However, there were some cassava wastes which could be used in the manufacturing of other consumers' products. Though many people do not consider the numerous benefits from cassava rhizome, and one of these benefits is the amount of energy that could be generated, which was found out to be an equivalent of single firewood [3]. In fact, it is classified as a biomass, for its unique potential for electricity generation.

Generally, the cassava plant is composed of root, top/leaf, stem, and rhizome Cassava roots are processed into pellets, chips and starch, mainly used as raw materials for several industrial and commercial purposes such as food, beverage, sweetener, textile, paper, glue and ethanol [4]. There are many routes to convert cassava waste for energy [5-10]. Biomass is accounted for a major fraction in renewable energy in Thailand. The eastern region of Thailand has a total area of 22.81x10^6 rai, which is agricultural cultivated area approximately 12.88x10^6 rai, representing 56.5 percent of the total Eastern region. The cultivated area for cassava is 1.5 x10^6 rai [11]. There are plenty of cassava residues, especially its rhizome, left unused in the field after harvesting. Thus, they show potential for use as a raw material in energy production. In addition, it was reported that the power plants in this region are facing a problem on raw material in terms of uncertainty and insufficient amount. Thus, cassava rhizome is promising, using as alternative raw material for the power plant. However, to promote the use of cassava rhizome it is important to know the fuel characteristics, potential along with its distribution in the region.

In this research, the main objective is to study potential of cassava rhizome using as raw material for energy production in terms of quantity and quality. The results can explore the suitability of cassava rhizome for energy production in the eastern region of Thailand.

2 Methodology

Our study on the potential generation of electricity from cassava rhizomes in the eastern region of Thailand is divided into 2 sections, which are: The survey and analysis of the quantitative data (quantity), and qualitative (quality) of the cassava stalks as a renewable energy through biomass.

* Corresponding author: Nanya_Sangsaior@gmail.com

2.1 Data and Survey

The survey and analysis of quantitative data on cassava rhizomes as a form of renewable energy focus on the followings:
- Information on cassava, and its uses for potential and quantitative data analysis.
- The natural quality of cassava rhizome in the eastern Thailand.
- And Information on cassava planting area, at the district and provincial levels of Thailand's eastern region.
- Information from the Agricultural Economic Development Agency on the average yield of cassava per ton per rai, of each province.
- Spatial data sets on the four-dimensional extent and scope of government (Sharp file and Digital file) from the Department of Land Development.

2.2 Data analysis

Quantitative data analysis (quantity) of cassava rhizomes for the study of spatial energy potential:
- Biomass Yield Ratio Crop Residue Ratios: (CRR) is the amount of biomass for energy generation without residue.
- Energy potential of biomass (kiloton per oil equivalent; toe).
- Details on area, quantity of existing biomass raw materials, and the estimation of potential to commerce acquisition for biomass feedstock.
- Data analysis on the quality of cassava rhizomes:
- Physical properties of biomass fuel such as: size, size distribution, moisture content, and appearance density.
- Proximate analysis of biomass fuel such as: moisture, ash, volatile matter, and fixed carbon. Analysis was performed according to ASTM E 1617-94 [12] and ASTM E 1755-01 [13], while according to ASTM D 2015 [14], a bomb calorimeter was adopted for different heating values.
- The ultimate analysis of biomass fuels such as: C, H, N, S, and O was conducted, moreover, elemental analyses of carbon (C), hydrogen (H), and nitrogen (N) were done with the use of a PerkinElmer 2400 CHNS Elemental Analyzer. The sulfur (S) content was determined by the use of a plasma spectrometry, while that of optical analysis was carried with a Sector ICP-OES spectrometer as established by titration according to [15].

3 Results

3.1 Quantity properties (quantity) of cassava rhizomes

The cassava is cultivated in the tropics to produce roots, used primarily for human consumption but also for animal feed and the extraction of starch. Cassava plants have a high biomass yield under good environmental conditions and a great ability to adapt to environments with biotic and abiotic stresses, such as poor-fertility or water-stressed soils where other cultures are not feasible. The aerial part of the plant consists of thick and thin stalks, petioles and leaves, and the underground part consists of roots and the seed stem and cassava stalks that were cut into approximately 200 mm-long sections and planted to reproduce the plant asexually, and after it grows it looks like woody biomass. The structure of the plant is shown in Fig. 1. Harvesting usually involves cutting the stalks, uprooting the plant from the soil and removing the roots. While the first two operations can be done by mechanical equipment, the seed stem and roots are separated manually as there is no mechanical equipment available to carry out this operation. After all the process of harvest, about half of the biomass produced is abandoned in the fields.

The cassava cultivated area in the eastern region of Thailand is 1,529,576.24, and the average value of fresh cassava rhizomes is about 359 kg/ rai. In addition, the ratio of cassava rhizome per (CRR) is approximately 0.12 as displayed in Table 1, while the number of the remaining fresh cassava rhizomes amounts to 636,003 tons/year with humidity of about 53.40% wb, and heating value of 12.50 MJ/kg as shown in Table 2.

Table 1. Crop residual ratio: CRR

Biomass	Fresh cassava weight (kg/rai)	Fresh cassava rhizomes weight (kg/rai)	Crop residual ratio: (CRR)
cassava rhizomes	8,757,938.2	636,003.35	0.112

Fig.1. Cassava rhizomes

Table 2. Residual cassava rhizome in the eastern region at Thailand

province	1*Average yield (Ton/ Year)	Cultivation area 2* (Km^2)	Cultivation area 3* (rai)	4*Cassava product (Ton/Year)	5*crop residual ratio:CRR	6*Cassava rhizomes (Ton/Year)
Chanthaburi	3.24	125,013,295.53	78,133.31	253,151.92	0.112	30,125.08
Rayong	4.40	130,864,317.17	81,790.20	359,876.87	0.112	42,825.35
Chon Buri	4.02	400,684,547.12	250,427.84	1,007,471.21	0.112	119,889.07
Sa Kaeo	3.29	965,895,178.60	603,684.49	1,986,926.87	0.112	236,444.30

Chachoengsao	3.38	395,434,852.24	247,146.78	835,603.27	0.112	99,436.79
Prachin Buri	3.36	429,419,720.39	268387.32	901,513.03	0.112	107,280.05
Tart	3.62	10,068.14	6.29	22.75	0.112	2.71
Total	-	-	1,529,576.24	253,151.92	-	636,003.35

Table 3. Residual cassava rhizome in the eastern region at Thailand

province	Cultivation area (rai)	Average yield (Ton/ Year)	Cassava rhizomes (CRR)
Chanthaburi	78,133.31	3.24	30,125.08
Rayong	81,790.20	4.40	42,825.35
Chon Buri	250,427.84	4.02	119,889.07
Sa Kaeo	603,684.49	3.29	236,444.30
Chachoengsao	247,146.78	3.38	99,436.79
Prachin Buri	268387.32	3.36	107,280.05
Tart	6.29	3.62	2.71
Total	1,529,576.24	-	636,003.35

The results of our study as shown in Table 2 and Table 3 reveal the prospect of electricity generation from cassava cultivation through biomass processing technique in the eastern region of Thailand. This tendency for energy generation is publicized and substantiated by the statistical results of 1,529,576.24 rai cassava cultivated areas of the eastern's seven provinces. An average yield of 3.59 tons of cassava per rai, and the remaining cassava rhizomes are 636,003.35 tons per year. Furthermore, the density's range of cassava stalks as a potential raw material for biomass is divided into 5: density fewer than 1,000 tons, density range of 1000-3000 ton, 3000-5000 tons, 5,000 - 10,000 tons per year, and a density of more than 10,000 tons per year. The research on cassava stalks in the eastern region is presented in Figure 2, whereas, the spatial density of cassava rhizomes as displayed in Figure 3 reveals the richness of Wang province over its neighborhood., Sa Kaeo Province and Chachoengsao province (red zone) have more than 10,000 tons of cassava rhizomes per year.

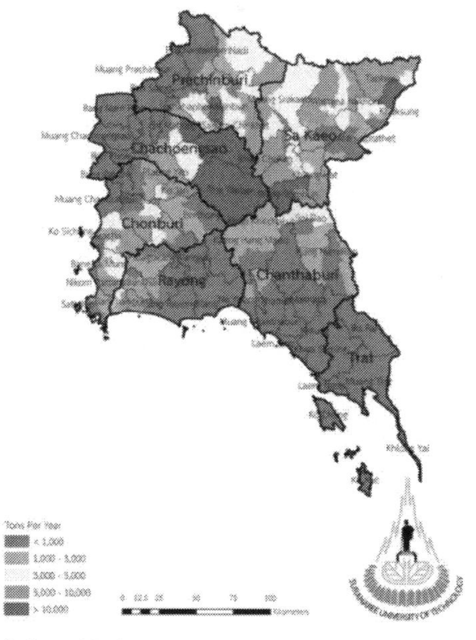

Fig.2. Potential of cassava rhizomes in the eastern region

Fig. 3. The potential for energy production

3.2 Quality properties (quality) of cassava rhizomes

The sample analysis on the quality of cassava rhizomes revealed that the qualitative and qualitative analyses of cassava roots could be used in the production of renewable energy. The physical properties of biomass fuel such as: size, size distribution, moisture content), appearance densities were also duly investigated and examined. Lastly, the Proximate and Ultimate analyses were carried out.

The physical properties of cassava rhizomes such as; size, density, and moisture content of (40-70%) are presented in Table 4. These irregular sizes and shapes depend on species, and different techniques of harvesting (machine and human labor). Most biomass fuels are of low-densities with replica features; nonetheless, there are some exceptions in the commercialized biomass fuels.

Table 4. Crop residual ratio: CRR

Types of biomass	moisture (%wb)	density (kg /m^3)	generalization
Cassava rhizome	53.40	193	Sectional diameter of 1-2 inches of the stem end.

An exclusive property of biomass elements, which is the Proximate Value was highly observed and thoroughly scrutinized in our study. Though, the ratio of Moisture Content, amount of Volatile Matter of Fixed Carbon, Ash Content and the ASTM standard method of analysis in the selection process were appropriately considered. The heat experiment adopted the use of cassava roots as the main fuel in the analysis of heat, and humidity relationship. From the analysis of biomass moisture, it was concluded that the heating value of cassava rhizomes as an alternate source of fuel was inversely proportional to the cassava roots' humidity. It was discovered that the cassava roots of moisture content 53.40% has a low calorific value (LHV) of 8.19 MJ/kg, this value serves as a threshold value for generating electricity or heat energy. Nevertheless, if the moisture content of the cassava rhizomes has humidity value of 4.66%, the heating value of fuel would be significantly by 14.591 MJ/kg as shown in the Table. 5 and also Table 6.

Table 5 Chemical properties of cassava rhizomes

Proximate Value					
Moisture Content (%)	Proximate Value			Heating Value	
	Volatile Matter (%)	Ash Content (%)	Fixed Carbon (%)	Volatile Matter (%)	Ash Content (%)
4.66	80.00	2.73	13.00	8,190	14,591
Ultimate Value					
The proportion of the elements (percent).					
C	H		N	S	O
46.12	7.55		1.13	0.03	54.83

Table 5, depicts the information of our proposed source of energy from the agricultural wastes of Thai's cassava plantations. Their thermochemical properties, proximate, ultimate, structural, inorganic matter, heating values, and thermo gravimetric analyses are determined, and rigorously examined in this study. Our results illustrate that these wastes have high volatile contents (of 80%, on dry basis), and contain 46.12% of carbon that is slightly different from those of [16] ,which portrays C content of about 51% ,1.13% hydrogen ,54.83 % of oxygen, and a biomass high heating value (HHV) of approximately 14.159 MJkg^{-1}.

3.3 The potential for energy production

In addition, our results indicate that cassava wastes possess high volatility tendency and very low nitrogen with a medium heating value. Outcomes form our study indicate that an environmentally friendly and nonthreatening bio-fuel could be produced via thermochemical conversion process that is identical to that of fast pyrolysis.

Likewise, Dioxins could be formed from all combustion processes, when organic carbon and oxygen are present [17], and these Dioxins are considered to be very hazardous to human health. Moreover, an acid rain is a severe problem, due to the burning of fuel with high sulphur content. But cassava wastes are environmentally friendly, due to its low S content of about 0.3-0.2% respectively. The use of cassava wastes prevents the emission of dioxins, HCl, SOx, and other harmful particles as stated in [18]. Hot corrosion of super heaters could also be enhanced by different surfaces, it increases the foul smells from the heat transfer of chlorine due to the formation of chloride, and sulphates, which vaporize and condense on cold surfaces later on [19, 20].The small amounts of S and Cl embedded in the ash are caused by the high contents of alkali and earth-alkali metals. Therefore, reducing the quantities of S and Cl will completely eradicate the problems of acid rain [18].

Table 6. The potential for energy production

province	Cassava rhizome (Ton/Year)	Loss rate		heating Value MJ/kg	Potential energy					
		Dehumidification 80% (Ton/Year)	Manufacturing 30% (Ton/Year)		Biomass Energy potoentail (GJ/Year)	Energy potoentail (Ton of Oil Equivalent, toe)	Dehumidification 80% (Ton/Year)	Manufacturing 30% (Ton/Year)	Power Plant capacity (MW)	Power Plant capacity (kW)

Chanthaburi	30,125.08	6,025.02	4,217.51	12.50	52,718.89	1,259.11	2,928,827.11	2,928.83	0.418	417.93
Rayong	42,825.35	8,565.07	5,995.55	12.50	74,944.36	1,789.93	4,163,575.48	4,163.58	0.594	594.12
Chon Buri	119.889.07	23,977.81	16,784.47	12.50	209.805.88	5,010.89	11,655,882.17	11,655.88	1.663	1,663.23
Sa Kaeo	236,444.30	47,288.86	33,102.20	12.50	413,777.52	9,882.43	22,987,640.08	22,987.64	3.280	3,280.20
Chachoengsao	99,436.79	19,887.36	13,921.15	12.50	174,014.38	4,156.06	9,667,465.63	9.667.47	1.379	1,379.49
Prachin Buri	107,280.05	21,456.01	15,019.21	12.50	187,740.09	4,483.88	10,430,004.86	10,430.00	1.488	1,488.30
Tart	2.71	0.54	0.38	12.50	4.74	0.11	263.22	0.26	0.000	0.04

The cultivated area for cassava plantation in the Eastern region of Thailand is 2,447,321,979 Square kilometers or 1,529,576 acres, and it has yielded an average of 3.62 tons per acre over the past three years. The ratio of the cassava stalks production to biomass is at 0.112 of the 636,003 tons of cassava rhizomes planted per year in the Eastern region. Likewise, cassava stalks with the heating value of 14,591 kilojoules per kilogram (kJ/kg) of the whole eastern rhizome wastes, could generate 9.27 x 10^9 MJ per year.

However, cassava stalks' production for fuel availability is highly essential, and also the dehumidifying of moisture content from 53.40% to 10.17%, which represents an 80% weight loss according to the (Department of Energy, 2557). This loss that occurs from the method of collecting, processing, chopping, and drying of bars is equivalent to 30% (Department of Energy, 2557). The amount of biomass residues that could be used as fuel for electricity generation is given in Table 5. The potential energy production values at Sa Kaeo, Chachoengsao, Chon Buri, Rayong, Chanthaburi, Prachin Buri and Trat provinces are: 52,718.89 74,944.36 209,805.88 413,777.52 174.01. 187,740.09 4.38, and 4.74 MJ per year, equivalent of a renewable energy (oil) of 1,259.11 1,789.93 5,010.89 9,882.43 4,156.06 4,483.88 0.11 toe.

4 Conclusion

There have been different biomass processing techniques for cassava wastes in recent years. However, our research centres on the analysis of a suitable processing biomass technique for generating electricity from cassava wastes in the eastern region of Thailand. Our case study focuses on the eastern region of Thailand, because it is the largest cassava cultivated area in Thailand. Our proposed technique and methods have produced some remarkable and outstanding results, which are highlighted in this section. Crop residual ratio (CRR) of cassava rhizome was 0.112 with the remaining amount of 636,003.35 tons/year (~53.40% wb moisture content and 8.19 MJ/kg heating value). The estimated amount of cassava rhizome with chip/pellet grade of moisture content value of ~10.17% wb, and heating value of ~12.50 MJ/Kg, is 89,040 tons/year, though the equivalent energy is 1,113.01 GJ or 26.58 ktoe. These methodical results have shown that 61.8 GWh of electricity could be generated to drive a plant of 8.8 MW.

The use of cassava wastes as an innovative and potential source of energy has opened up an interesting area of research. However, some measures and precautions should be prudently observed in order to avoid an unlikely occurrence of incrustation, due to the harmful characteristics of ashes from these wastes. In addition, a drying process technique must be adequately considered for the high moisture content present at the time of harvest, before these wastes are used for fuel production. Furthermore, a thorough evaluation of cassava production in other regions of Thailand is significantly necessary. Our study has provided right information on the agronomic, chemical, and physical characteristics of cassava wastes for energy generation. Wood, sugarcane bagasse, and ethanol represent different sources of energy production and consumption in Thailand; nevertheless, other modern forms of biomass should be encouraged and promoted [21]. The advancement of other forms will enhance a rapid development of sugarcane sector, and many industrial sectors with different sources of agricultural wastes for energy production and sustainability. Thereby reducing the level of uncertainty, that is associated with fossil-fuel market.

A rapid development of cassava wastes would strengthen its feasibility as another form of raw material for an integrated electricity generation, ethanol, and food production. This technique has been adopted by and used in the sugarcane sector of Brazil.

Our study has revealed and reinforced the versatile use of cassava crop for food, ethanol, and thermal energy. However, our study focuses on the promotion and use of rhizomes as a commercial generated biomass fuel. Though, there are other non-renewable fuels such as: crude oil and coal, however, cassava rhizomes serve as an alternative source of energy. In order words, it will help to limit the frequent shortage of commercial and industrial biomass fuels for the production of electricity and heat. Lastly, our future work will focus on how to enhance other potential agricultural wastes for energy generation

References

1. B. Ekasingh" C. Sungkapitux, J. Kitchaicharoen, and P. Suebpongsang, *Competitive Commercial Agriculture in the Northeast of Thailand*. A World Bank Research Publication, (2007)
2. Ministry of Commercem The export production from export structure of Thailand, (2009). [Online] Avaliable: http://www2.ops3.moc.go.thiexport!recode export! [20 July 2010]
3. Thailand Biomass Potential Study Report 2010, Department of Alternative Energy Development
4. Prakobboon N, Vahdati M. Review of the potential for co-firing of cassava rhizome for generating heat and power in cassava-based bio-ethanol plant in Thailand. International Journal of Biomass and Renewables ,**2**(2):14-22. (2013)
5. Punnarapong P, Promwungkwa A, Tippayawong N. *Development and performance evaluation of a biomass*

gasification system for ceramic firing process. Energy Procedia ,**110** :53-58. (2017)

6. Jaroenkhasemmeesuk C, Tippayawong N. *Technical and economic analysis of a biomass pyrolysis plant*. Energy Procedia ,**79**: 950-955. (2015)
7. Tippayawong KY, Tippayawong N. Fuel recovery from thermal processing of post-consumer footwear waste. Energy Engineering,**114**(3):7-16 (2017)
8. Wongsiriamnuay T, Tippayawong N. Effect of densification parameters on property of maize residue pellets. Biosystems Engineering ,**139**:111-120 (2015)
9. Malek K, de la Seigliere C, Fernandez C, Swaminathan M, Sebastian A, Arora D. Green coal: *a new energy source from leaves*. Energy Procedia,**100** :484-491. (2016)
10. Pattiya A, Titiloye JO, Bridgwater A. Evaluation of catalytic pyrolysis of cassava rhizome by principal component analysis. Fuel,**89**:244-253. (2010)
11. *The report on the agricultural economy of Thailand in 2017*, Agricultural Information Center, Agricultural Economics Office
12. E1617-09 *Standard Practice for Reporting Particle Size Characterization Data*. Book of Standards Volume: 14.02.
13. ASTM E1755-01-*Standard Test Method for Ash in Biomass*. Book of Standards Volume: **11**.06.
14. ASTM D2015-00 *Standard Test Method for Gross Calorific Value of Coal and Coke by the Adiabatic Bomb Calorimeter*.
15. APHA, American Public Health Association, Standard Methods for the Exam- ination of Water and Wastewater, Method 4500-Cl B, twentieth ed., Water Environment Federation, American Water Works Association, Washington DC, USA, (1999)
16. A. Pattyia, Thermochemical characterization of agricultural wastes from Thai cassava plantations, Energy Sources, Part A Recover. Util. Environ. Eff. **33** (8) (2011) 691e701, http://dx.doi.org/10.1080/15567030903228922.
17. E.D. Lavric, A.A. Konnov, J. De Ruyck, Dioxin levels in wood combustionda review, Biomass Bioenergy **26** (2) (2004) 115e145, http://dx.doi.org/10.1016/S0961-9534(03)00104-1.
18. I. Obernberger, T. Brunner, G. B€arnthaler, Chemical properties of solid bio-fuels significance and impact, Biomass Bioenergy **30** (11) (2006) 973-982 http://dx.doi.org/10.1016/j.biombioe.2006.06.011.
19. S. Arvelakis, H. Gehrmann, M. Beckman, E. Koukios, Studying the ash behav-iour of agricultural residues using thermal analysis, J. Therm. Analysis Calorim. **72** (2003) 1019-1030.
20. T.R. Miles, J.R. Miles, L.L. Baxter, R.W. Bryers, B.M. Jenkins, L.L. Oden, Boilerde posits from firing biomass fuels, Biomass Bioenergy **10** (2) (1996) 125e138, http://dx.doi.org/10.1016/0961-9534(95)00067-4.
21. J. Goldemberg, S.T. Coelho, Renewable energydtraditional biomass vs. mod- ern biomass, Energy Policy **32** (6) :711-714, (2004)

E3S Web of Conferences **182**, 02001 (2020) https://doi.org/10.1051/e3sconf/202018202001
CPEEE 2020

Performance of HSDI diesel-engine generator using the blend of B5, n-butanol and ethanol as increased to 20%

Ekkachai Sutheerasak[1,*], *Worachest* Pirompugd[1], and *Wirogana* Ruengphrathuengsuka[2]

[1]Department of Mechanical Engineering, Faculty of Engineering, Burapha University, Chonburi, Thailand
[2]Department of Chemical Engineering, Faculty of Engineering, Burapha University, Chonburi, Thailand

Abstract. B5, diesel mixed with 5% biodiesel, is currently being developed to replace diesel, but there was lower engine performance. To improve the B5 properties, the addition of oxygenated additive is a better method. This research aims to study the performance of a high-speed direct injection (HSDI) diesel-engine generator at speed 3,000 rpm and different loads by using B5 blended to n-butanol and ethanol as increased to 20%. Results show that the use of B5-butanol-ethanol blends decreased engine performance as increasing ethanol; however, the release of nitric oxide, carbon monoxide, and black smoke was remarkably reduced as compared with B5. However, the use of B5 blended to 5% n-butanol, and 5% ethanol increased the electrical power to 0.33%, while electrical efficiency was added to 1.13%, and SFC was similar to B5. Therefore, this ratio can be applied with the diesel engines in the future.

1 Introduction

Electricity is mainly generated from a power plant and sometimes producing from an engine-generator station for agriculture, facility, subsistence, etc. Energy sources are reducing continuously, resulting in the extremely price of primary fuels. Moreover, the innumerable release of carbon dioxide (CO_2) and particulate matter, particularly black smoke which had a diameter of less than 2.5 microns (PM2.5), from the electrical stations, especially diesel-engine generators, is increasing continuously and causing the destruction of the environment and human health. Ethanol and biodiesel become an alternative fuel attractive in blending with diesel since oxygen (O_2) content within both oils could decrease the diesel-engine emissions [1, 2].

Prior studies on the diesel-biodiesel-ethanol blends were indicating that the mixture of diesel, biodiesel, and ethanol by using emulsification is the best because there was highly homogeneous as depended on proportion, purity, temperature, and solubility [2-4]. Ethanol could not be mixed more than 10% because of stratification time happening quickly [4-6]. And the use of diesel blending to biodiesel and anhydrous ethanol, which should not more than 10%, was lower engine power than diesel but the exhaust gas emissions, especially carbon monoxide (CO) and black smoke were decreased [7-15].

To increase the homogeneity, some researches [2-5] used emulsifiers, such as n-butanol, ethyl acetate, propanol, etc., in blending to diesel and ethanol. Use of 5% butanol, which was normal butanol or n-butanol, mixed with diesel and anhydrous ethanol could improve engine performance, and diesel could be blended to 25% anhydrous ethanol by high homogeneity, and lower thermal efficiency leading to the increase of fuel consumption [5]. Prior studies on the diesel-biodiesel-anhydrous ethanol blends by adding n-butanol for HSDI diesel engines had some parts. B5, which was diesel mixed with 5% palm methyl ester (PME), is currently replaced the conventional diesel in Thailand. Still, this research has developed the B5 as produced from using 5% palm ethyl ester (PEE) to replace the use of PME in the future. This research aims to study the fuel properties and performance of an HSDI diesel-engine generator using the B5, the 95%w diesel blended to 5%w PEE, mixed with n-butanol and anhydrous ethanol as increased to 20% comparing to B5 by testing at a constant speed 3,000 rpm and loads.

2 Methodology

2.1 Fuel preparation

Substances consist of D, B5, anhydrous ethanol (99.9%w), and n-butanol (99.5%w). For the sequence of the mixture, n-butanol (Bu) was used at 5%w in all proportions, and B5 was blended to ethanol (E) by increasing to 5, 10, 15, and 20%w, while terms were shown as B5E5Bu5, B5E10Bu5, B5E15Bu5, and B5E20Bu5 respectively. In the emulsion process, the electromagnetic machine and the mechanical stirrer were applied in the controlling mixture at 800 rpm, and the blending temperature was fixed at 30 °C as studied from the phase diagram in Reference [4]. Next, there were investigating physical properties, such as fuel density, kinematic viscosity, and lower heating value (LHV), under various ASTM procedures, as shown in Table 1.

* Corresponding author: ekkachai@eng.buu.ac.th

Table 1. Fuel properties.

Items	Density (kg/m^3)	Viscosity (cst)	LHV (MJ/kg)	Separated time
ASTM	D1298	D445	D240	-
D	830	3.26	44.57	-
B5	836	3.32	43.40	Not
B5E5Bu5	809	3.06	42.03	Not
B5E10Bu5	803	3.01	41.87	25 (days)
B5E15Bu5	797	2.78	40.99	16 (days)
B5E20Bu5	790	2.55	39.98	11 (days)

Fuel properties' results of B5 blended to ethanol from 5 to 20%w and 5%w n-butanol as compared with B5 under various ASTM procedures showing that fuel density was decreased from 3.23 to 5.50%, kinematic viscosity was reduced from 7.83 to 23.19%, and LHV was decreased from 3.15 to 7.88%. Moreover, this research found that B5E5Bu5 had not been separated for 2 months. As compared with Reference [1], B5E5Bu5 could be applied as an alternative fuel with diesel engines in the future because some properties, especially viscosity and density, were within the scope of standard diesel as determined by the Department of Energy Business, Thailand.

2.2 Engine performance test

The experiments were carried out on an HSDI diesel-engine generator [Model, Mitsuki: 5GF-ME; cylinder, 1 cyl; capacity, 0.406 L; power (max.), 5 kW$_{ele}$ @ 3,000 rpm; compression ratio, 17.5:1], while the schematic of the experimental setup is shown in Fig. 1.

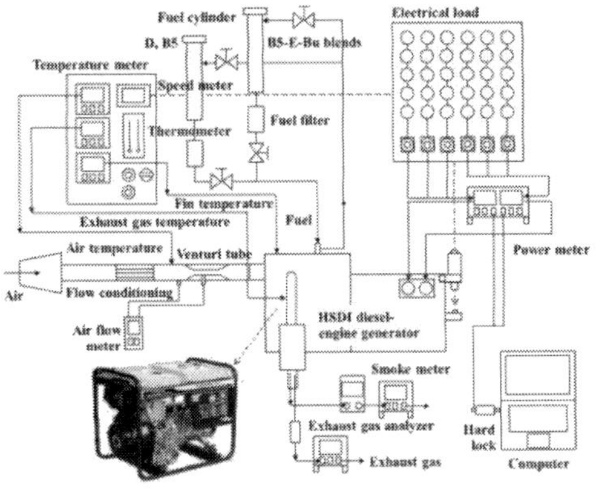

Fig. 1. Schematic of the experimental setup.

This generator was connected with the light-bulb panel, which had several bulbs, for increasing electrical loads. At the same time, a power meter was applied in measuring the electrical power by processing on a computer. Engine temperatures, such as coolant, air, and exhaust gas, were investigated by using the K-type thermocouple and the temperature meters. The air flow rate was measured from a venturi tube and an air flow meter. Fuel consumption was calculated from the amount of fuel 20 ml within a fuel cylinder per time. Finally, there were the installation of Cosber: KWQ-5 automotive emission analyzer and Cosber: KYD-6 opacimeter to measure the CO and NO levels and the black-smoke intensity, respectively.

For investigating the engine performance from using these oils, this generator was firstly warmed up about 15 minutes. After engine operation was stable, the surrounding temperature was controlled at 30±3 °C, and the coolant temperature was set at 90±5 °C. All experiments were started up by using diesel (D). Speed was controlled at 3,000±50 rpm. Electrical loads were started from 20% and later 40%, 60%, 80% and 100% respectively. Fuel volume was fixed at 20 ml to record the change of time. Parameters, such as electrical power, air flow rate, fuel consumption, temperatures, and emissions, were recorded. Next, the B5 and the B5-ethanol-butanol blends (B5-E-Bu blends), such as B5E5Bu5, B5E10Bu5, B5E15Bu5, and B5E20Bu5, were tested respectively, as using the same condition with diesel test. Finally, all parameters from using D, B5, and B5-E-Bu blends were calculated in the engine-performance parameters. Particularly, thermal efficiency as calculated in case of electrical efficiency, which was the ratio of electrical power to the sum of fuel consumption and calorific value, and specific fuel consumption as studied from the ratio of fuel consumption to electrical power. All period of the engine test was between 100 hours, and results were repeated by more than 5 times [3].

3 Results and discussion

3.1 Electrical power

Fig. 2 indicates that the electrical power is increased as increasing loads, while the use of B5-E-Bu blends at the electrical load from 20 to 80% gave the electrical power similar to D and B5.

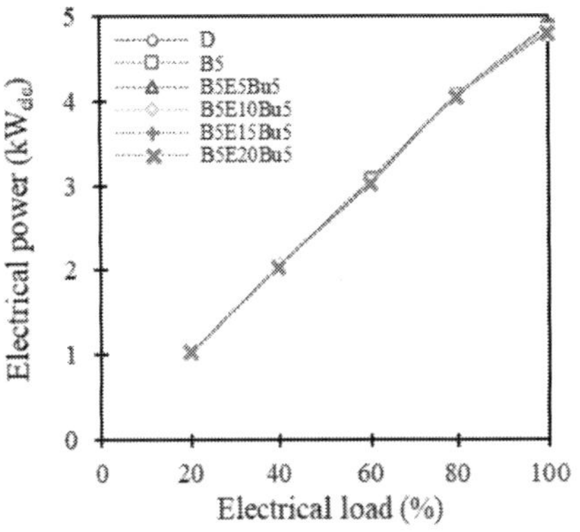

Fig. 2. Electrical power at different loads.

However, the use of B5-E-Bu blends at full load (100%), showing that electrical power as generated from using these oils changing. Especially, B5 mixed with 5% n-butanol and 20% ethanol (B5E20Bu5) had lower electrical power than B5 as reduced to 1.75%. This result is consistent with literature reviews [1-6]. Because B5 was typically higher heating value than ethanol, ethanol, as added to 20% mixing with B5 and n-butanol, resulted in a highly lower calorific value than B5 (Table 1), leading to the decrease of electrical power. Contrarily, this research is focusing on the change of electrical power from using B5 mixed with 5% n-butanol and 5% ethanol (B5E5Bu5), as found that electrical power was increased to 0.33% as compared with B5. For adding the generating electricity, this research discusses that mixing 5% n-butanol with B5 and 5% ethanol resulted in the complete combustion leading to the addition of energy release in diesel-engine combustion processes. As a result, the net engine power was increased, leading to the addition of output engine power as converted into electrical power.

3.2 Electrical efficiency

Fig. 3 on the left side shows that electrical efficiency (EE) is increased until 80% load, while the use of full load has a decrease of this efficiency due to increased heat loss and mechanical frictions [2]. This research found that two main points are interesting. First, B5E5Bu5 had higher EE than B5 as added to 1.13% at 80% load. This result is hypothesized by mixing 5% n-butanol with B5, and 5% ethanol led to a rapid ignition timing resulting in a reduced ignition delay. As a result, energy release in periods of diesel-engine combustion processes was increased, leading to an increase in thermal efficiency [4-6].

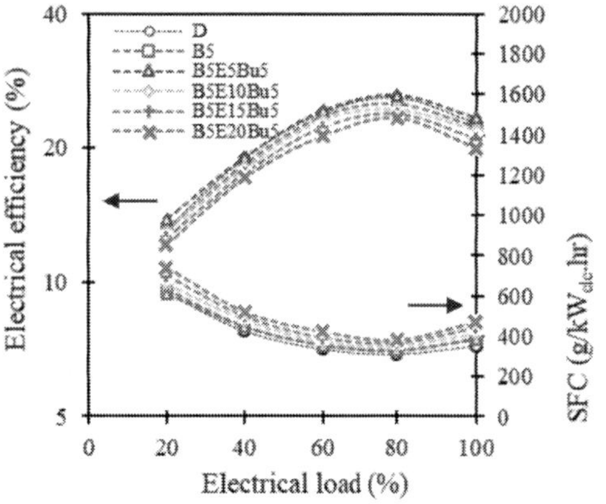

Fig. 3. Electrical efficiency and SFC.

Another reason is discussed by adding ethanol from 10 to 20% as mixed with B5 and n-butanol resulted in the decrease of EE, as reduced from 0.64 to 1.72% at 80% load compared with B5. Results were in the same direction of literature reviews [2-4] since n-butanol and ethanol had lower calorific value than B5 as blended to n-butanol and ethanol, adding from 10 to 20% (B5E10Bu5 to B5E20Bu5) resulted to the high reduction of heating value. As a result, the use of B5E10Bu5 to B5E20Bu5 had, respectively, higher fuel consumption and input fuel energy than B5 as producing electricity equally. For applying the B5-E-Bu blends as replaced to B5 with diesel-engine generators in the future, this research recommends that B5E5Bu5 is the best.

3.3 Specific fuel consumption

Fig. 3 on the right side shows that the specific fuel consumption (SFC) is decreased as increasing load until to 80%, and there is the increase of SFC at full load because of the rise in fuel consumption coming from the energy loss within the engine systems [2]. Similarly, the results of SFC have two interesting issues. First, SFC from using B5E5Bu5 was similar to B5 as hypothesized by B5E5Bu5 had the start of combustion rapidly leading to the decrease of ignition delay and complete combustion than B5. As a result, there was an increase in energy release in the combustion processes of the diesel engine. Although B5E5Bu5 had lower calorific value and higher fuel consumption than B5, they were only slightly [4-6]. For using B5E10Bu5 to B5E20Bu5, SFC was increased with the increase of ethanol as added from 6.32 to 16.47% at 80% load compared with B5. The results of this engine test had the same tendency as the previous researches [2-4]. Because the addition of ethanol from 10 to 20% as blending to B5 and n-butanol led to the decrease of heating value (Table 1) respectively, the electricity generation from using these oils equalled to using B5 resulting to higher fuel consumption.

3.4 Exhaust gas temperature

Fig. 4 on the left side shows the level of exhaust gas temperature (EGT) increasing with the increase of loads. For using the mixture of B5, n-butanol and ethanol adding from 5 to 20% as compared with B5, EGT was reduced from 13.98 to 32.95 °C at 80% load. These results have the opposite effect on References [2, 3].

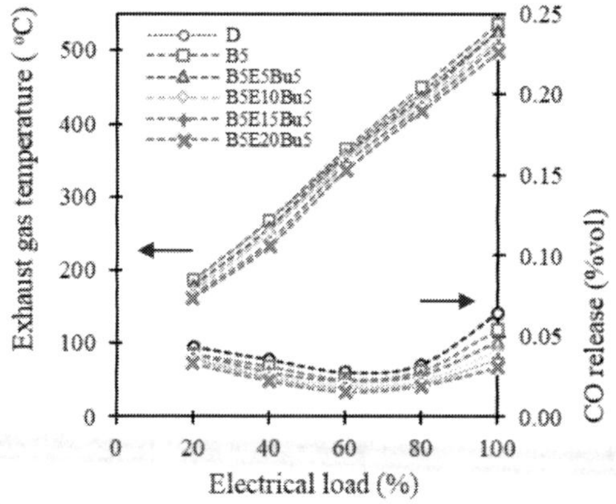

Fig. 4. Exhaust gas temperature and release of CO.

For the reasons of decreasing EGT from using these oils, firstly explained by this research used the primary substrate, which was B5 as produced from 95% diesel and 5% palm ethyl ester. Usually, the mixing ethyl ester with diesel led to the increase of burning temperature in premixed and mixing controlled combustion resulting in the increase of EGT as explained in [11-14]. Contrarily, the addition of n-butanol and ethanol up to 20% led to the change of diesel-engine combustion characteristics. Since ethanol and n-butanol had higher latent heat of vaporization than B5, B5 mixed with n-butanol and ethanol adding to 20% as injected into the combustion chamber resulted to the start of combustion quickly at a lower gas temperature which caused to the decrease of burning temperature in the cylinder. As a result, the burning temperature in mixing controlled combustion period was decreased [4-6]. Another reason is discussed by the addition of n-butanol and ethanol was to increase the O_2 content, resulting in the highly complete combustion in the premixed combustion phase leading to the duration of this period increased. As a result, the duration of mixing controlled combustion was decreased leading to the end of combustion before the end of power stroke while the exhaust valve was opened resulting to the decrease of EGT [4-6].

3.5 Carbon monoxide release

Fig. 4 on the right side indicates that the increase of load from 20 to 60% leads to the reduction of carbon monoxide (CO) emission, while CO level is increased as adding load from 80 to 100% because of increasing electricity generation causing to the increase of fuel consumption and the decrease of air-fuel ratio [2]. Use of B5-E-Bu blends comparing with B5 showing that CO release was decreased as increasing loads and ethanol percentages. For investigating at 80% load, CO level was reduced from 8.77 to 31.58% while these results were in the same direction of literature reviews [4-6] because of adding ethanol causing to the reduction of ignition delay from burning reactant by using high O_2 content resulting to the increasing products, particularly CO_2. Moreover, the use of B5 has higher CO release than the use of B5 mixed with n-butanol and ethanol as increased from 5 to 20%, because the B5 was produced from the 5% palm ethyl ester resulting to the increase of burning rate in mixing controlled combustion phase leading to the addition of CO level [11-14]. Therefore, the increase of O_2 content from mixing n-butanol and ethanol, which increasing until 20%, with B5 resulted in the complete combustion. O_2 element within these oils was mainly reacted with one carbon atom within these oils becoming to CO_2 [4-6].

3.6 Nitric oxide release

Fig. 5 on the left side shows that release of nitric oxide (NO) is increased, as the electrical load is added from 20 to 80%. However, the level of NO is decreased at full load because there was an increase of CO level as explained in Section 3.5. Use of B5 blended to n-butanol and ethanol as increased from 5 to 20% resulted in the decrease of NO level, as the electrical load was increased. For measuring the NO release at 80% load, NO level was reduced from 1.85 to 17.50% as compared with B5. These results are consistent with previous researches [4-6] while two interesting issues are illustrating the decrease of NO emission.

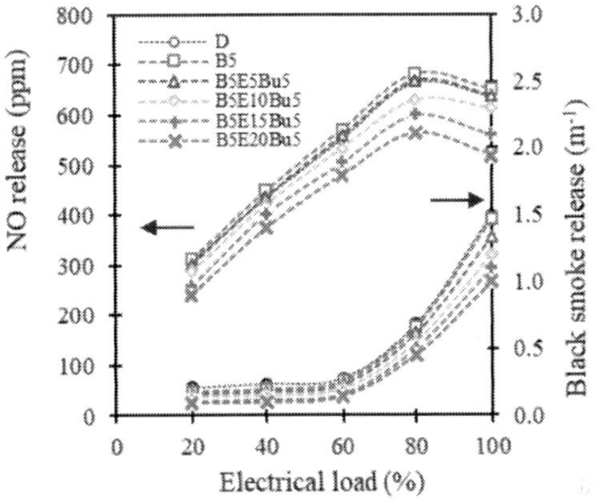

Fig. 5. NO level and release of black smoke.

First, ethanol and n-butanol had higher latent heat of vaporization after being injected into the combustion chamber, causing the start of combustion quickly at a lower gas temperature in the cylinder. As a result, the burning temperature in periods of diesel-engine combustion processes was decreased, resulting in less NO formation. In addition, ethanol and n-butanol were lower calorific value than B5 leading to the reduction of the heating value of the B5-E-Bu blends, which caused lower combustion temperature than B5. Another reason is explained from NO formation came from increasing the oxygen concentration in during the combustion processes of diesel engine, but the addition of ethanol concentration as mixing with B5 and n-butanol led to the complete combustion. As a result, O_2 element within these oils was mainly reacted with one carbon atom becoming to CO_2 as discussed in Section 3.5.

3.7 Black smoke release

Fig. 5 on the right indicates that the intensity of black smoke is increased as adding load, while the use of B5-E-Bu blends leads to the decrease of black-smoke intensity as reduced from 7.06 to 31.92% at 80% load compared with B5. Results of black-smoke intensity could conclude that the use of B5 mixed with n-butanol and ethanol as adding to 20% reduced the black smoke release, while the results of this engine test have the same tendency as previous researches [2-6]. Reason of black smoke level was decreased as using the B5-E-Bu blends, firstly confirmed from decreasing of CO level as shown in Section 3.5. Next, the use of B5 as compared with diesel resulted in the level of black smoke decreased because B5 had the O_2 element leading to complete combustion in mixing controlled combustion period [11-15]. For mixing n-butanol and ethanol, as increasing to 20%, with diesel, literature reviews [4-6] ensures that there was the increase of O_2 content leading to the ratio of carbon to hydrogen (C/H ratio) of D-E-Bu blends decreased resulting to the reduction of black smoke formation in the exhaust gas. Therefore, the use of B5 blended to n-butanol and ethanol as adding to 20% leading to the lower C/H ratio and higher O_2 content than B5. High O_2 element from using the B5-E-Bu blends had improved better combustion efficiency in mixing controlled combustion phase resulting in a higher reduction of black smoke release than B5.

4 Conclusions

From investigating the fuel properties and performance of HSDI diesel-engine generator by using B5-E-Bu blends as compared with B5 at 3,000 rpm and different loads. The main results are concluded in the following items :

(1) Physical properties of B5-E-Bu blends were lower than B5, but B5E5Bu5 had some properties, particularly viscosity and density, which was within the scope of standard diesel as determined by the Energy Business Department, Thailand. B5E5Bu5 had higher homogeneity. For applying as an alternative fuel with the diesel engines in the future, this research suggests that B5E5Bu5 should be used.

(2) Electricity generation from using the B5-E-Bu blends is similar to B5 while the electrical load was increased from 20 to 80%, but the ability to generate electricity at full load showing clearly that the use of B5E20Bu5 had lower electrical power. Contrarily, the use of B5E5Bu5 gave higher electrical power than B5.

(3) Electrical efficiency from using B5E5Bu5 was higher than B5 to 1.13%, but the use of B5E20Bu5 was decreased to 1.72%. Contrarily, SFC from using B5E5Bu5 was similar to B5, but the use of B5E20Bu5 resulted in an increase of SFC highly. Therefore, the choice of the right mixture between B5, n-butanol and ethanol as the replacement to B5 in the future recommends that B5E5Bu5 is the best because of better engine performance.

(4) Results of exhaust-gas emissions shows that the use of B5 mixed with n-butanol and increasing ethanol up to 20% as compared with B5 led to the decrease of exhaust gas temperature and pollutants, particularly CO, NO, and black smoke levels. Reasons resulting to the reduction of pollutants are explained by ethanol and n-butanol had higher latent heat of vaporization, lower C/H ratio, and higher O_2 content leading to the decrease of pollutants.

(5) The results of the present study provide the ideas for further study improvement. The injection and combustion characteristics and other emissions, such as CO_2 and unburned hydrocarbon, from using the B5-E-Bu blends as compared with B5 must be studied. While this investigating indicating that B5E5Bu5 was the best, the study of the wears of the engine from the use of B5E5Bu5 as compared with B5 in the long term should be studied.

The authors wish to thank N. Prasangsasakul, P. Sabsiri and I. Sabkoon for collected experimental data. This work was supported by the Faculty of Engineering, Burapha University, Thailand (Grant No. 169/2562) is gratefully acknowledged.

References

1. S.M. Krishna, P.A. Salam, M. Tongroon, N. Chollacoop, App. Ther. Eng. **155**, 9 (2019)
2. E. Sutheerasak, U. Ratch, Uni. Eng. J. **10**,14 (2017)
3. E. Sutheerasak, C. Chinwanitcharoen, U. Ratch. Uni. Eng. J. **9**, 13 (2016)
4. I. Yahuza, H. Dandakouta, Chem. Eng. Pro. Tech. **6**, 6 (2015)
5. S. Kumar, J.H. Cho, J. Park, I. Moon, Ren. Sus. Ene. R. **22**, 26 (2013)
6. R. Niculescu, A. Clenci, V. Iorga-Siman, Ene. **12**, 41 (2019)
7. A. Jamrozik, W. Tutak, M. Pyrc, M. Sobiepanski, Ther. Sci. **21**, 14 (2017)
8. S. Shamun, G. Belgiorno, G. Di Blasio, C. Beatrice, M. Tunér, P.Tunestål, App. Ther. Eng. **145**, 9 (2018)
9. A.O. Emiroğlu, M. Şen, App. Ther. Eng. **133**, 10 (2018)
10. F. Pradelle, S.L. Braga, A.R. Fonseca de Aguiar Martins, F. Turkovics, R.N.C. Pradelle, Ren. Ene. 136, 13 (2019)
11. H. Tse, C.W. Leung, C.S. Cheung, Adv. Auto. Eng. **1**, 7 (2016)
12. D. Alviso, F. Krauch, R. Román, H. Maldonado,R.G.D. Santos, J.C. Rolón, N. Darabiha, F. **191**, 16 (2017)
13. R.I. Yahuza, Ejilah, H. Dandakouta, S.S. Farinwata, Bio. Pro. O. A. J. **2**, 9 (2018)
14. U. Vats, R. Chaudhary, S. Kumar, A. Giri, S. Kumar, Inter. J. App. Eng. R. **13**, 7 (2018)
15. S. Shinde, D.D. Palande, Inter. R. J. Eng. Tech. **3**, 4 (2016)

Application of natural language understanding in Chinese power dispatching centre

Jiahong Tong[1,*], *Zhigang* Wu[1], *Qi* Liu[2], *Liang* Du[2], *Liangde* Xu[2] and *Jingxing* Yu[3]

[1] School of Electrical Engineering, South China University of technology, Guangzhou 510640, China
[2] Power Dispatching and Controlling Center, Guangzhou Power Supply Co., Ltd., Guangzhou 510620, China
[3] Reading Academy, Nanjing University of Information Science and Technology, Nanjing 210044, China

Abstract. It is difficult for computer to understand the texts in unstructured Chinese language, which becomes an obstacle for further application of artificial intelligence in the power dispatch center. Understanding of the orders from human dispatchers is the premise for the collaboration of machine and human being in power system operation. Towards understanding of dispatching texts, this paper proposes a textual semantic analysis framework with active learning of the semantic structure knowledge. Firstly, the words are vectorized by the Skip-gram models. And the hierarchical clustering algorithm is designed to detect the sentence patterns. Then the knowledge base is set up by converting the sentence structure to their regular expressions. In application, define a proprietary semantic framework to extract important device information and to parse the semantic slot using dependency syntax. Application shows that the Chinese texts describing the operation mode switching process can be understood accurately by the computer program.

1 Introduction

The role of artificial intelligence in the power system is gradually evolved from computer-aided to computer-partner in some tasks. In this process, human-machine collaboration is essential. In power dispatching center, a typical application scenario is that a human expert set up a plan for the operation mode change in the next day. An intelligent robot takes over the task to check out the security of the plan and to carry out the switch process. In this human-computer collaboration link, an important technical obstacle to is how to enable the computer to correctly understand the professional text content written in natural language so as to translate them into the machine data formats.

Although artificial intelligence technology represented by deep neural networks has achieved great success in the fields of speech recognition and language translation, understanding the information expressed by natural language and matching them with real world objectives and operations are still hot technical issues to be studied. Natural Language Understanding (NLU) requires the knowledge of natural language syntactic, professional domain knowledge and vocabulary, and a variety of machine learning techniques. In the application of power dispatching, the target of NLU to extract the information of objects, actions and sequence from the natural language expression. Combination of the NLU technology with the big data analysis and other artificial intelligence technology can promote a broad prospects in human-machine coordination of power systems.

A main task in power dispatching center is to adjust the power system structure and operation modes according to maintenance schedule or accidents. Currently many tasks, such as power flow analysis, breaker remote switching and adjustment of the protection settings, are already carried out through computers. However, organization of the tasks and process is still relied on human operators. In China, developing of the machine dispatcher becomes a hot spot for power grid enterprises in the past two years. NLU will be a kernel link in such machine dispatcher because human-machine collaboration as well as communication in Chinese natural language is inevitable. For instance, the computer dispatcher need to read the operation plan provided by professional technicians and check its security before execution. Currently, this plan is written by professional technicians in texts of natural language. The operation mode documents cover a wide range of professional fields and include a large number of professional terms, certain normative expression requirements. Anyway it still retains the arbitrariness of Chinese natural language expression. It is a challenge to develop targeted and efficient techniques to achieve the NLU of such power dispatch texts.

This paper proposes a set of semantic analysis algorithms in specific to the power dispatching text which can learn the semantic expression knowledge independently. For the learning of phase, the text dictionary is mined using the Bi-LSTM+CRF model, and

* Corresponding author: 2448778655@qq.com

the Skip-gram neural network is used to implement dictionary vectorization and statement vector representation. The pattern clustering of sentence vector structure is realized based on hierarchical clustering algorithm (HCA). Furthermore, the longest common subsequence algorithm is used to extract the regular expression specification, which constitutes the prior knowledge base of the semantic analysis model. In the application phase of semantic analysis, Populate the semantic slot using dependency syntax analysis techniques. The effectiveness of the proposed model and algorithm is verified in practical application.

2 Pretreatment

2.1 Named Entity Recognition (NER) based on Bi-LSTM+CRF

There are a large number of long and complete nouns in the power dispatch plan text (DPT), which will greatly reduce the correct rate of Chinese word segmentation. Therefore, it is necessary to use NER technology to discover power proper nouns. In this paper, the Bi-LSTM+CRF model[1] is used. The schematic diagram of the model is shown in Figure 1 There are a large number of long and complete nouns in the power dispatch plan text (DPT), which will greatly reduce the correct rate of Chinese word segmentation. Therefore, it is necessary to use NER technology to discover power proper nouns. In this paper, the Bi-LSTM+CRF model[1] is used. The schematic diagram of the model is shown in Figure 1.

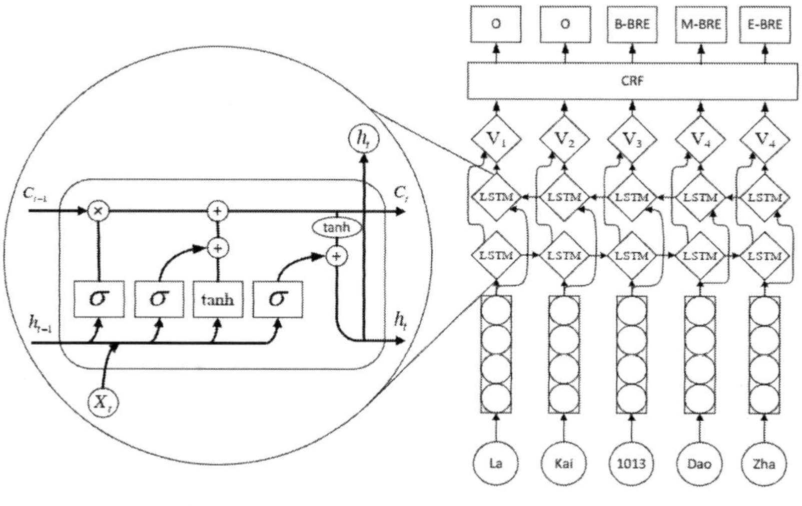

Fig. 1. LSTM neuron structure and Bi-LSTM+CRF structure.

In Figure 1, B represents the first character of the word, M represents the middle character of the word, and E represents the last character of the word.

The LSTM neuron calculation formula is as shown in formulas (1)-(6):

$$f_t = \sigma(W_f \cdot [h_{t-1}, x_t] + b_f) \quad (1)$$

$$i_t = \sigma(W_i \cdot [h_{t-1}, x_t] + b_i) \quad (2)$$

$$\tilde{C} = \tanh(W_C \cdot [h_{t-1}, x_t] + b_C) \quad (3)$$

$$C_t = f_t * C_{t-1} + i_t * \tilde{C}_{t-1} \quad (4)$$

$$o_t = \sigma(W_o[h_{t-1}, x_t] + b_o) \quad (5)$$

$$h_t = o_t * \tanh(C_t) \quad (6)$$

Where x_t is the embedding of a word which is the input of current neuron, we use One-hot in this paper. h_{t-1} is the output of the previous moment. σ is sigmoid function.

This method outputs the entity state of each character. Besides, ordinary part-of-speech tagging is rough and cannot meet the needs of grid text. For example, the names of equipment such as transformers, plant stations, switches, etc. are all marked as nouns, and understanding the grid text requires obtaining more subtle part of speech, that is, entity part of speech. Therefore, we design a part-of-speech suffix to the entity tag. For detailed suffix meaning comments, see the appendix.

2.2 Segment

Chinese word segmentation technology has been relatively mature. After obtaining the power entity dictionary after 2.1, the common open source word segmentation tool can meet the correct rate requirement. This article selects the Jieba word segmentation tool[2] in Python.

3 Sentence classification based on Skip-gram model and hierarchical clustering

3.1 Skip-gram

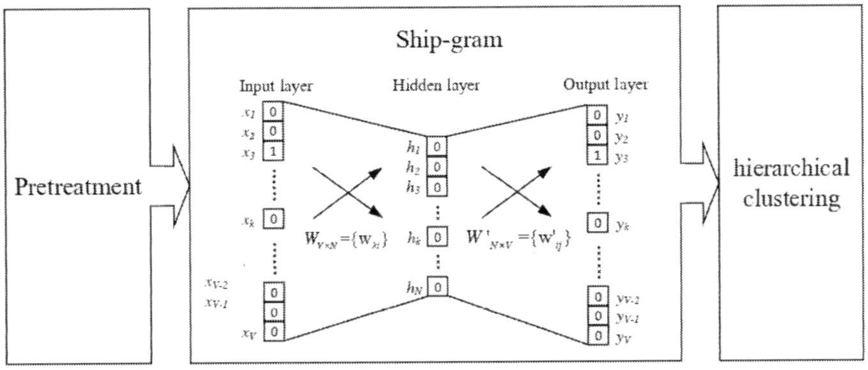

Fig. 2. Skip-gram model and hierarchical clustering.

Applying statistical analysis algorithms to natural language texts, we need to transform the texts into structured data at first. To this end, a word-embedding model can be used to convert the words in the power dispatch dictionary established in the previous section into vectors of fixed length, and it is expected that the distance between the vectors can reflect the correlation between words. Common representative word embedding models are: CBOW (Continuous Bag Of Words) and Skip-gram [3]. Figure 2 shows the algorithm flow chart.

In this paper, we apply the above two models to vectorize the words of the scheduled text samples. Then their application performance is compared by subsequent clustering effects. In our applications, the Skip-gram model show better performance than the CBOW model. According to the scale of the scheduling dictionary, the length of the word vector selected in this paper is 100 dimensions.

For each actual DPT statement, the arithmetic mean of the word vector is taken as a sentence vector for clustering the scheduled statement samples. We implement the above functions based on the Word2Vec module in the gensim toolkit[4] of the Python programming language.

3.2 Sentence summary based on the longest common subsequence (LCS)

After finding the samples with similar sentence structure by clustering algorithm, this paper uses the LCS algorithm to extract the common part words of the same sample to form the sentence rules of the category. Figure 3 shows the calculation process of the LCS algorithm through an example. Where x and y are two operational sentences in the same cluster, and the two-dimensional matrix C is calculated according to formula (7).

$$C[i,j]=\begin{cases}0 & if\ \ i=0\ or\ j=0\\ C[i-1,j-1]+1 & if\ i,j{>}0 \text{ and } x_i=y_j\\ \max(C[i,j-1],C[i-1,j]) if\ i,j>0 and x_i\neq y_j\end{cases}$$

The regular expression paradigm is used to describe the expression specification as well as the requirements for the variable part. Repeating LCS algorithm on all the sentences in the same cluster, we can obtain a general expression of a class of scheduling sentences, and derive its regular expression(RE), as shown in Table 1. Finally the regular expression knowledge base for the power dispatch plan text can be established autonomously.

Table 1. Regular Expression examples Using LCS Algorithm

Examples	LCS	RE
The #2 transformer height switch of Chengbei station is switched to 1M bus for operation. (城北站#2 变高开关倒至 1M 母线运行)	X zhan X bian X kai guan X zhi X mu xian yun xing	^[\u4e00-\u9fa5]{0, 3}zhan[\u4e00-\u9fa5#0-9]{0, 5}bian[\u4e00-\u9fa5]{0, 3}开关(dao)?(hui fu)?zhi[A-Za-z0-9]mu xian yun xing$
The #1 transformer height switch of Chengbei station is switched to 2M bus for operation. (城北站#1 变高开关恢复至 2M 母线运行)		
The transformer height switch of #2 main transformer of Chengbei station is switched to 110kv1m bus for operation (城北站#2 主变变高开关倒至 110kV1M 母线运行)		

4 Key word extraction

4.1 Semantic framework

Each DPT sentence describes a specific operation in a certain operation process, such as state checking and break switching. This paper uses the semantic framework to define the expression and storage structure of keywords such as substations and equipment. The corresponding text parsing task is a process of correctly identifying the operation object described by each DPT sentence and correctly filling the keyword information into the corresponding attribute of the semantic framework. For example, Table 2 shows the semantic framework of the bus charging operation sentence. Each DPT sentence describes a specific operation in a certain operation process, such as state checking and break switching. This paper uses the semantic framework to define the expression and storage structure of keywords such as substations and equipment. The corresponding text parsing task is a process of correctly identifying the operation object described by each DPT sentence and correctly filling the keyword information into the corresponding attribute of the semantic framework. For example, Table 2 shows the semantic framework of the bus charging operation sentence.

Table 2. Semantic framework definition of bus-to-power operation type

Layer 1	Power outage equipment			Power supply equipment		
Layer 2	Station	Voltage	Bus	Station	Voltage	Bus
	**zhan	220kV	6M	-	-	5M

4.2 Dependency Parsing Analysis (DPA)

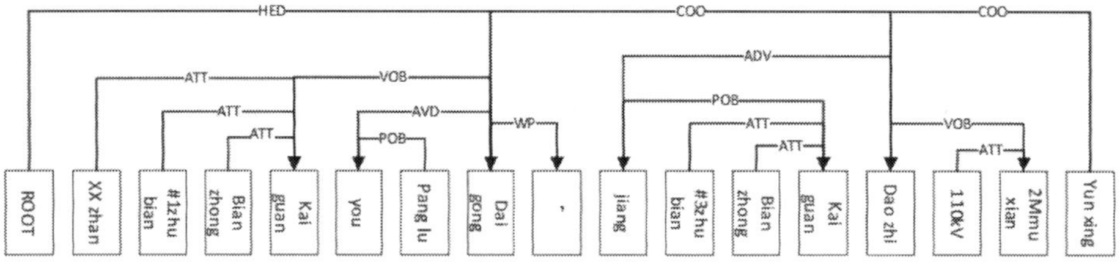

Figure 3. An Example of DPA

The premise of DPA is that only one independent component is existed in a sentence which does not depend on any other component. DPA can obtain a syntactic analysis model through learning the co-occurrence relationship and the dependence of words from DPT samples. Figure 3 is an example of syntactic analysis using DPA. This work is based on the Python package of the LTP language platform[5,6]. The detailed meaning of the word dependency in the figure is shown in Appendix A.2.

5 Test result and discussion

5.1 Learning stage results

A total of 8813 DPT samples from January 2017 to June 2018 were obtained from a dispatching center. After the clauses saperation, more than 30,000 operational sentences were generated to form the learning set.

Apply the new word discovery algorithm proposed in this paper, 24777 words are extracted to form the professional dictionary of DPT. Among these words, there are 674 transformer related, 732 line related, 5210 breaker related, 949 bus related, 15340 switch related, 684 substation related and 88 generator related words. Other words covers name of departments, protection and automation devices etc.

Figure 4 show the results of text vectorization and hierarchical clustering in three graphs. The left picture shows the sentence vector using the t-SNE algorithm to reduce the dimension, and the middle picture is the tree diagram of the hierarchical clustering of the sentence vector. In this paper, the total of clusters is selected as 38. The graph on the right shows the clustering results for each of the 38 categories. This work is based on the Scipy toolkit and implementation in python.

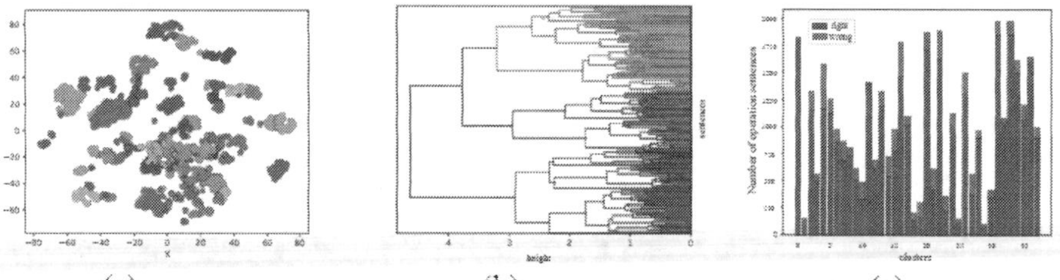

(a) (b) (c)

Figure 4. Use t-SNE [7] dimension reduction sentence vector (a) and hierarchical clustering result (b) and Cluster correct rate(c)

5.2 Testing stage results

We use the DPT of the dispatching center from July 20 to December 31 in year 2018 to form the test set. for the trained NLU model. There were totally 2,141 dispatching plans in the set. All the clauses in test set are identified using the regular expression knowledge base obtained by the technique in 3.2, and the identification correctness rate is shown by the confusion matrix of Figure 5.

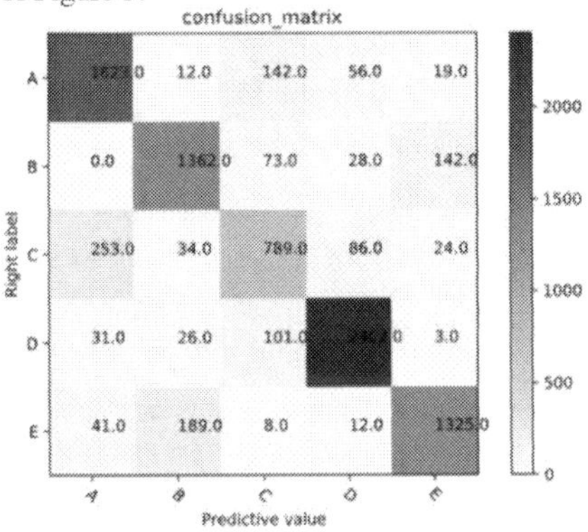

Figure 5. Confusion Matrix

The semantic resolution rate is shown in Table 3. Here we defines two types of error indices: general errors(GE) and important errors(IE). The general error refers to the incorrect padding information of the non-essential semantic slot. Such errors can be found and correctly located to the operating device by matching with the records in the power supply device database. A serious error means that the important semantic slot information is incorrect, which may result in the operating device being unable to locate or locate the error.

Table 3. Semantic Parsing Error Rate

	A	B	C	D	E
GE	1.20%	0.51%	1.12%	1.34%	0.90%
IE	0.52%	0.21%	0.85%	0.92%	0.90%

6 Conclusion

In this paper, a set of NLU combination algorithm is proposed for the power system operation schedule texts based on sample learning. The proposed method has the following characteristics. (1) Use the BiLSTM+CRF model to explore professional dictionaries and improve the accuracy of Chinese word segmentation. (2) Using hierarchical clustering to achieve syntactic pattern clustering based on large sample set. Furthermore, the LCS algorithm is used to autonomously generate the regular expression rules for each sentence cluster, which can form a regular grammar library. Based on this regular grammar library, the sentence recognition shows high accuracy and can support effective fuzzy recognition. (3) Establishing semantic frameworks to locate and describe different types of power equipment for the complex association of terminology in the power industry. The semantic slot is filled using the DPA method.

The research work is supported by the Guangzhou Power Suppl y Co. of China under Grant no. GZHKJXM20170059

Appendices

Table A1. Power word suffix

Power word tag	suffix	example
Station	S	XX zhan
Power Plant	PP	XX dianchang
Voltage	V	220kV
Bus	B	2M muxian
Line	L	XX yixian
Transformor	T	#1 zhubian
Breaker	BRE	#1 daozha

References

1. H. Zhiheng, X. Wei and Y. Kai, arXiv. preprint arXiv:1508.01991. (2015)
2. S. Junyi, Jieba. Onine. Available: https://github.com/fxsjy/jieba
3. T. Mikolov, I. Sutskever, K. Chen, G. Corrado and J. Dean, Conf. *Advances in Neural Information Processing Systems*; 26:3111-3119. (2013)
4. R. Řehůřek, P. Sojka, Conf. *Proceedings of the LREC 2010 Workshop on New Challenges for NLP Frameworks*. (2010)
5. D. Chen, C. Manning, Conf. *Proceedings of the 2014 Conference on Empirical Methods in Natural Language Processing*, (2014).
6. W. Che, Z. Li, T. Liu, Conf. *23rd International Conference on Computational Linguistics*, 23-27. (2010)
7. L. van der Maaten, G. Hinton, J. Mach. Learn. Res. 9(2008)

Steady-state Operation Area of VSC-HVDC Converter Station Connecting Renewable Energy Cluster by Isolated Network

Wenyuan Xian[1,2], *Ran* Ding[3], *Ying* Qiao[2,*], *Zongxiang* Lu[2], *Shangqiang* Li[2], and *Hong* Lin[1]

[1]School of Electrical Engineering, Xinjiang University, Urumqi 830047, China
[2]State Key Lab of Control and Simulation of Power Systems and Generation Equipment (Department of Electrical Engineering, Tsinghua University), Haidian District, Beijing 100084, China
[3] State Grid Jibei Electric Power Company, Xicheng District, Beijing 100053, China

Abstract. When the large-scale renewable power island is connected to VSC-HVDC transmission system, we should figure out the steady-state operation area of VSC-HVDC converter station. Based on the equivalent model of renewable energy island, the constraints of VSC-HVDC converter station are analyzed, and a fast method for calculating steady-state operation area of converter station is presented in this paper. The influence of equivalent line impedance of renewable energy island, transformer ratio of converter station and voltage setting value of grid-connected point on steady-state operation area of converter station is analyzed. And the three-dimensional view of the steady-state operation area of the converter station under different grid-connected voltage is depicted.

1 Introduction

VSC-HVDC transmission technology has obvious advantages in clean energy grid-connected, island power supply, offshore platform power supply and other technical fields([1],[2]). In recent years, more and more large-scale offshore and onshore renewable energy clusters are transmitted to load centers through VSC-HVDC transmission[3]. By the analysis of existing VSC-HVDC projects, it is seen that VSC-HVDC transmission technology has the ability to deliver isolated network renewable energy cluster.

The renewable energy island is decoupled from the AC power grid when it is delivered through the VSC-HVDC transmission system. The AC side control mode of the converter station is constant voltage/frequency control, which provides voltage and frequency reference for the renewable energy AC island system. Converter station equals as a voltage source, providing reactive power and voltage support for AC islands[4]. In such a system, the steady-state operation of converter station is one of the important factors that restrict the maximum power of renewable energy island[5]. Therefore, Research on constraints and influencing factors of steady-state operation area of converter station is a great significance about the steady-state operation of the whole renewable energy island delivery system. At the same time, it can provide guidance for the optimization of reactive power and voltage in renewable energy island.

At present, some achievements have been made in the study of the steady-state operation area of VSC-HVDC transmission system. The stable operation region of VSC-HVDC system connected with weak AC system is studied, and the stability constraints under various operation modes are given[6]. [7] analyzed the influence of AC network strength on the steady-state characteristics of VSC-HVDC transmission system. [8] studied the influence of short circuit ratio of AC system, shunt reactive compensation capacitor, converter transformer capacity and converter station capacity on stable operation range of converter station. But the particularity of the system is not considered. In practical engineering application, [9] proposed that the steady-state operation area of the converter station can be calculated under the maximum and minimum AC voltage in Zhoushan multi-terminal DC(VSC-MTDC) project, taking into account the tap position and the range of DC voltage. [10] analyzed the operation characteristics of the renewable energy isolated network connected to the VSC-HVDC power grid, considered the constraints that the steady-state operation interval should met, and describes the steady-state operation area by traversal method. And the influence of various operations on the steady-state operation area is analyzed, such as connecting transformer tap, the VSC-HVDC reference voltage, the impedance characteristics of equivalent distance of the renewable energy cluster and the VSC-HVDC converter station. However, traversal method needs to traverse all power points in unit circle of PQ power, which takes too long. And in the converter station, the voltage power operation characteristic curve changes from two-dimensional curve to three-dimensional space of P-Q-V. Therefore, it is inappropriate to measure the safety margin of converter station simply by using two-dimensional voltage-power characteristic curve.

* Corresponding author: qiaoying@tsinghua.edu.cn

https://doi.org/10.1051/e3sconf/202018202003

In summary, in view of the shortcomings of existing research, this paper aims at the renewable energy island network access to VSC-HVDC transmission system. Considering the internal parameter constraints of converter stations and the voltage constraints of renewable energy cluster, a fast method for calculating the steady-state operation region of converter stations is proposed. Finally, the influence of equivalent line impedance of renewable energy island, transformer ratio of converter station and voltage setting value of grid-connected point on steady-state operation area of converter station is analyzed. And the three-dimensional voltage-power operation characteristic curve is depicted, the influence of grid-connected voltage on steady-state operation area of converter station is more intuitively depicted.

2 System model

Fig. 1. shows the typical topology of renewable energy cluster integrated by islanded VSC-HVDC system. The radial network is composed of 220 kV AC. The renewable energy island accesses to the lower voltage level. There is no traditional synchronous unit in the renewable energy island. Each wind farm, photovoltaic power stations are directly assembled to the grid-connected point (PCC), and their outputs are then sent out through the converter station.

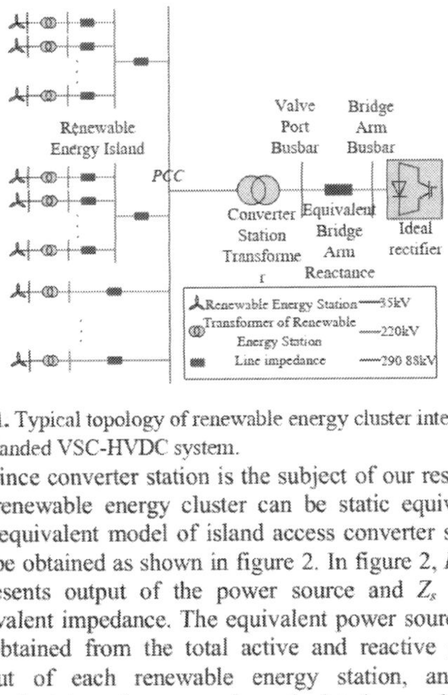

Fig. 1. Typical topology of renewable energy cluster integrated by islanded VSC-HVDC system.

Since converter station is the subject of our research, the renewable energy cluster can be static equivalent. The equivalent model of island access converter station can be obtained as shown in figure 2. In figure 2, P_s+jQ_s represents output of the power source and Z_s is the equivalent impedance. The equivalent power source can be obtained from the total active and reactive power output of each renewable energy station, and the equivalent impedance can be calculated according to equation (1).

In figure 2, $P_{cc}+jQ_{cc}$ is the power of grid-connected point, P_v+jQ_v is the power of converter station transformer, $P_\Delta+jQ_\Delta$ is the power of converter station, U_s is the voltage of equivalent power source, U_{pcc} is the voltage of grid-connected point, U_v is the voltage of converter station transformer, U_Δ is the voltage of converter station, X_T is the impedance of transformer in converter station, X_{L0} is the impedance of bridge arm in converter station, k is the variable Ratio of Transformer in Converter Station.

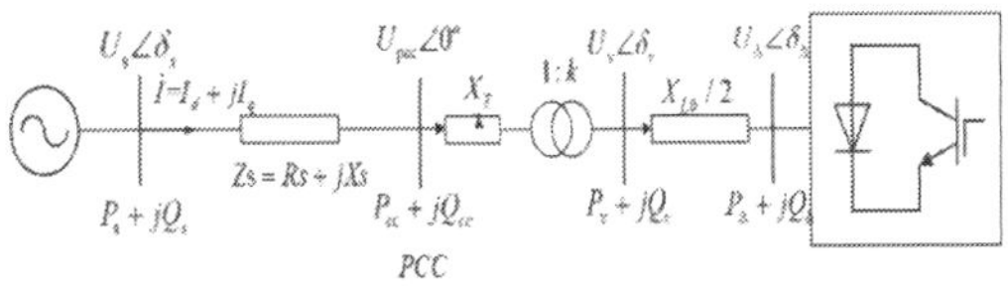

Fig. 2. Equivalent topology of renewable energy group integrated by islanded VSC-HVDC system.

$$Z_s = (P_{loss}+jQ_{loss})\left(\frac{\dot{U}_{pcc}}{P_{cc}+jQ_{cc}}\right)^*\left(\frac{\dot{U}_{pcc}}{P_{cc}+jQ_{cc}}\right) = (P_{loss}+jQ_{loss})\frac{U_{pcc}^2}{P_{cc}^2+Q_{cc}^2} \tag{1}$$

Where $(\cdot)^*$ denotes the conjugation. Under different output conditions of renewable energy cluster, the system losses are different, and so the equivalent impedance is also different.

3 Characterization method of steady-state operating area of converter station

3.1. Constraints of steady-state operating area of converter station

The steady-state operation area of converter station is a set of system operation points that satisfy the constraints on capacity of converter station. When the operation point of converter station is in the steady-state operation area, the system security can be guaranteed. The steady-state operation area of converter station is generally depicted in two-dimensional plane with active and reactive power in abscissa and longitudinal coordinates. Assume that the AC bus voltage U_{pcc} of a VSC-HVDC converter station is known at a certain time and the power flowing into the converter station is $P_{cc}+jQ_{cc}$, the operating range of the renewable energy VSC-HVDC system should satisfy the following constraints:

a. Capacity constraint of VSC-HVDC converter station

$$\sqrt{P_{cc}^2+Q_{cc}^2} \le S_{ccN} \tag{2}$$

where S_{ccN} is the rated capacity of converter station.

b. Capacity constraints of transformers in VSC-HVDC converter station

$$\sqrt{P_v^2+Q_v^2} \le S_{vN} \tag{3}$$

where S_{vN} is the rated capacity of transformer in converter station.

c. AC current constraint of VSC-HVDC converter station

$$\frac{\sqrt{P_{cc}^2+Q_{cc}^2}}{kU_{pcc}} \le I_{ccN} \tag{4}$$

where I_{ccN} is the rated current of converter station.

d. DC current constraint of VSC-HVDC converter station

$$-I_{dc\max} \le \frac{P_{cc}}{U_{dcN}} \le I_{dc\max} \tag{5}$$

where $I_{dc\max}$ is the maximum DC current of converter station, U_{dcN} is the DC side rated voltage.

e. Modulation ratio constraints of VSC-HVDC converters

$$m_{\min} \le m = \frac{2\sqrt{2}U_{\Delta}}{\mu U_{dc}} \le m_{\max} \tag{6}$$

where m is the modulation ratio of VSC-HVDC Converter, $m_{\max}$ is the maximum modulation ratio, $m_{\max}$ is the minimum modulation ratio, $m_{\max}$ equals 1.15. μ is the utilization rate of DC voltage, which equals 0.866. U_{dc} is the DC bus voltage.

f. Transformer ratio constraint in converter station

$$k_{\min} \le k \le k_{\max} \tag{7}$$

where k is the transformer ratio of converter station. $k_{\max}$ equals 1.075, $k_{\min}$ equals 0.925.

g. Voltage constraints of systems

$$U_{\min} \le U_j, U_{pcc}, U_s \le U_{\max} \tag{8}$$

where $U_{\min}$ is the minimum voltage constraints of systems, which equals 0.97. $U_{\max}$ is the maximum voltage constraints of systems, which equals 1.07.

h. Constraints on operation range of renewable energy stations

$$\begin{cases} Q_{s\min} \le Q_s \le Q_{s\max} \\ P_{s\min} \le P_s \le P_{s\max} \end{cases} \tag{9}$$

where Q_s, $Q_{s\min}$ and $Q_{s\max}$ are respectively the total reactive power output, the minimum total reactive power output and the maximum total reactive power output of the renewable energy station. P_s, $P_{s\min}$ and $P_{s\max}$ are respectively the minimum and maximum total effective output of renewable energy station.

i. Constraints of system power flow equation

$$\begin{cases} P_i = U_i \sum_{j=1}^{N} U_j (G_{ij}\cos\theta_{ij} + B_{ij}\sin\theta_{ij}) \\ Q_i = U_i \sum_{j=1}^{N} U_j (G_{ij}\sin\theta_{ij} - B_{ij}\cos\theta_{ij}) \end{cases} \tag{10}$$

where P_i and Q_i are the injection active power and reactive power of node i, U_i is the voltage value of node i, G_{ij} and B_{ij} are the branch conductance and branch admittance between node i and node j, and indicate the phase difference between node i and node j.

3.2 Description method of steady-state operating area of converter station

Obtaining the set of power points (P_{cc}, Q_{cc}) satisfying the constraint of formula (2) ~ (10) is the base for analysis on steady-state operating area. We can consider the constraint of formula (2) as a unit power circle and the steady-state operation area of the converter station must be within this unit circle. Therefore, we can find the power point satisfying the constraints from the boundary of the unit circle inward. The specific steps are as follows:

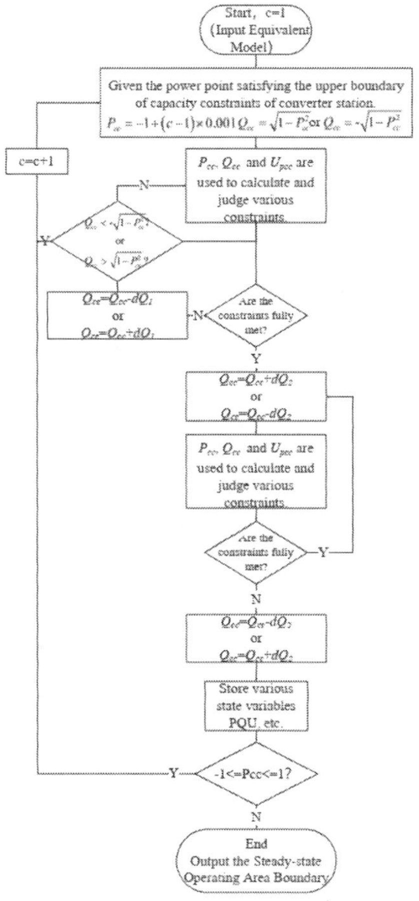

Fig. 3. Flow charts for describing the steady-state operation area of converter station.

Step 1. Input the equivalent model of the system.

Step 2. The power point satisfying the capacity constraint boundary of converter station is given.

Step 3. The electrical quantities in the system can be calculated by equation (10). Then each electrical quantity is substituted into equation (3) ~ (9) to determine whether all constraints are satisfied at the same time. If not, Q_{cc} subtracts step dQ_1 downward (Q_{cc} adds step dQ_1 up) (The length of dQ_1 should not be too large, otherwise it is possible to cross the steady-state operation area.), and judges whether the value of Q_{cc} exceeds the lower boundary of the unit circle (If it exceeds, go back to step 2, if it does not, go back to step

3.). If satisfied, it proves that the current power point has been in the steady-state operation area of the converter station, and the next step is taken.

Step 4. Because dQ_1 is set relatively large, the power point at this time may cross the boundary of safe and stable operation area, so Q_{cc} needs to increase (reduce) dQ_2 (dQ_1 is greater than dQ_2.) upward (down). The electrical quantities in the system can be calculated by equation (10). Then each electrical quantity is substituted into equation (3) ~ (9) to determine whether all constraints are satisfied at the same time.

Step 5. If the constraints are met, it means that the power point is still in the steady-state operation area, and return to step 4. If not, then Q_{cc} subtracts (adds) dQ_2, and stores various state variables corresponding to the current power point.

Step 6. Judge whether the value of P_{cc} is in the unit circle, if it is, return to step 2. If it is not, end the cycle, and describe the upper boundary of safe and stable operation area of converter station with all stored power points.

The steady-state operation area of the converter station can be obtained by the steps displayed in Fig. 3. It should be noted that in order to improve the calculation speed, we do not use ergodic method, and we rather set the corresponding step size dQ_1 and dQ_2. The set of power points satisfying constraints can be found more quickly by changing step size.

4 Case study

The shape, size and location of the steady-state operation area of the converter station are affected by the constraints shown in formula (2). For a certain renewable energy island VSC-HVDC delivery system, most of its system parameters are fixed, such as the total power output limit and the modulation ratio of converter station, and so forth. However, the equivalent line impedance of renewable energy islands, the voltage setting value of converter station and the converter conversion ratio may vary during the operation of the system, thus changing the size of other electrical quantities of the system. Therefore, the sensitivity analysis method is used to explore the influence of these three factors on the shape, size and relative position of the steady-state operation area of the system. The unit circle is the capacity constraint of the converter station.

4.1 Equivalent line impedance of renewable energy island

According to equation (1), the equivalent line impedance of renewable energy island is related to the total power output of renewable energy station. Fig. 4. shows the steady-state operation area diagram of the grid-connected point under different equivalent impedance. It can be found that with the increase of equivalent impedance, the steady-state operation area decreases gradually. This shows that the magnitude of equivalent impedance has a great influence on the steady-state operation area.

In fact, for a certain renewable energy island system (just like the system mentioned in this paper), the equivalent line impedance will not change much with the power output of the station. As shown in Fig. 5. When the output power of all renewable energy station changes from 0% to 100%, the maximum change of equivalent impedance resistance is not more than 0.0001, and the maximum change of reactance is not more than 0.003.

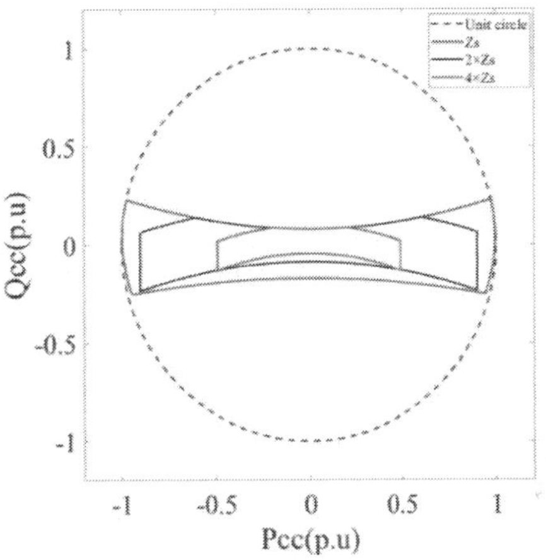

Fig. 4. Steady-state operation area of converter station under different equivalent impedances.

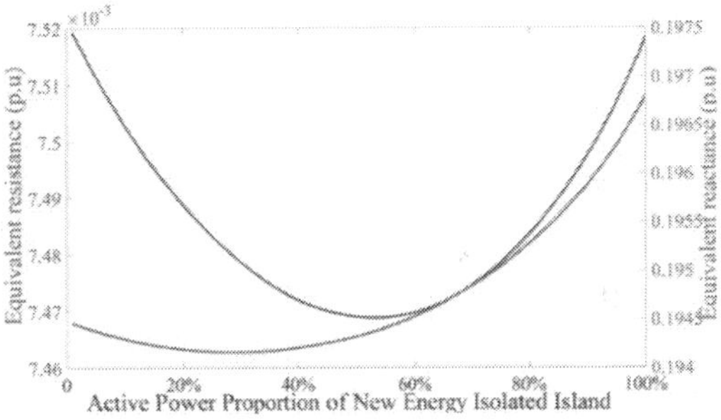

Fig. 5. Renewable energy island equivalent impedance value with different active power output ratio.

4.2 Transformer ratio of converter station

Under fixed islanding equivalent impedance and grid-connected voltage setting, changing transformer ratio of converter station, the steady-state operation area of converter station under different situations can be obtained, as shown in Fig. 6.

With the change of transformer ratio in converter station, its steady-state operation area will change for both shape and size. The reason is that when the transformer ratio of converter station changes, the reactive power distribution of the system will change, so the position of steady-state operation area will shift up and down. It can be seen that the setting of transformer ratio in converter station is also very important for the steady-state operation of converter station.

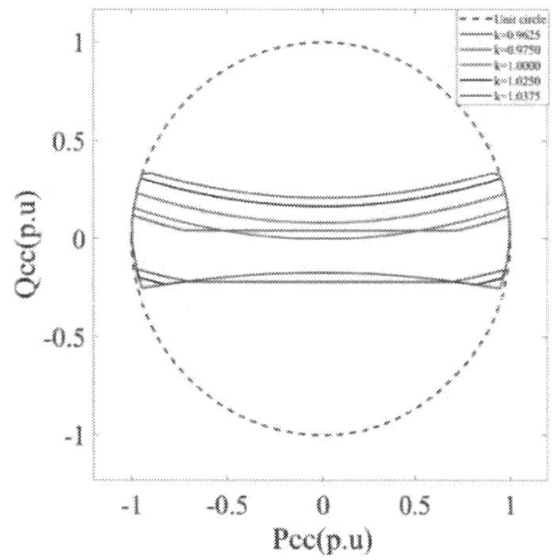

Fig. 6. Steady-state operation area of converter station under different transformer ratio of converter station.

4.3 Voltage setting value of grid-connected point

Under fixed islanding equivalent impedance and transformer ratio of converter station, we changed grid-connected voltage setting, and the steady-state operation area of converter station under different situations can be obtained, as shown in Fig. 7.

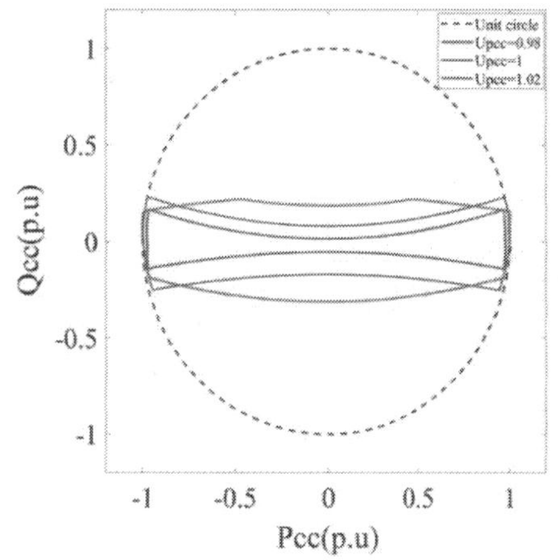

Fig. 7. Steady-state operation area of converter station under different grid-connected voltage settings.

With the increase of voltage setting value, the shape of steady-state region changes obviously. Fig. 8 shows a three-dimensional view of the steady-state operation area of the converter station under different grid-connected voltages. The influence of grid-connected voltage on steady-state operation area can be seen more intuitively from the figure. When the voltage is at a certain value, the maximum steady-state operation area can be obtained. With this voltage as the starting point, increasing or decreasing the voltage will reduce the steady-state operation area.

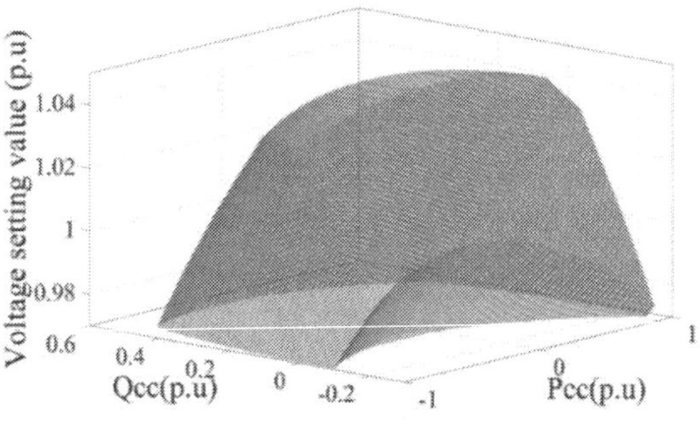

Fig. 8. Three-dimensional view of the steady-state operation area of the converter station under different grid-connected voltage.

5 Discussion and conclusion

In this paper, we study on steady-state operation area of VSC-HVDC converter station connecting renewable energy cluster by isolated network. The conclusions are as follows:

Based on the analysis on the constraints for the steady-state operation area of the converter station, a method is proposed to characterize the steady-state operation area of the converter station. This method can depict the steady-state operation area of converter station more quickly.

Although the magnitude of equivalent impedance has an influence on the steady-state operation area, for a certain renewable energy island system, the equivalent line impedance will not change much with the power output of the station. Therefore, for a certain renewable energy island system, only the transformer ratio of converter station and voltage setting value of grid-connected point have great influence on the steady-state operation area of converter station.

From the three-dimensional diagram, we can see the influence of grid connected voltage on the steady-state operation area more intuitively.

This work was supported by State Grid Corporation Science (SGJB0000TKJS1801210).

References

1. Kang, H. S. , and A. Yazdani . [IEEE 2016 North American Power Symposium (NAPS) - Denver, CO, USA (2016.9.18-2016.9.20)] 2016 North American Power Symposium (NAPS) - Enhancement of power system stability with VSC-HVDC transmission. *North American Power Symposium* IEEE, 2016:1-5.
2. Guangfu Tang, Zhiyuan He, Hui Pang. Research, Application and Development of VSC-HVDC Engineering Technology. *Automation of Electric Power Systems*, 2013, 37(15): 3-14.
3. Korompili, Asimenia, Q. Wu, and H. Zhao. Review of VSC HVDC connection for offshore wind power

https://doi.org/10.1051/e3sconf/202018202003

integration. *Renewable and Sustainable Energy Reviews*, 59(2016):1405-1414.

4. Yongning Chi, Zhankui Zhang, Yan Li, et al. Development of Large-scale Wind Power Grid Integration and Technical Standard. *North China Electric Power*, 2017, (3):1-7.
5. Linlin Wu, Man Xu, Hui Liu, et al. Optimal Voltage Control Strategy for Islanded Aggregated New Energy System Connecting by VSC-HVDC. *Journal of Global Energy Interconnection*, 2018,1(02):112-119.
6. Sheng Liu, Zheng XU. Study on Stable Operating Region of VSC-HVDC Connected to Weak AC Systems. *Proceedings of the CSEE*, 2016,36(01):133-144.
7. Zhou J Z, Gole A M. VSC transmission limitations imposed by ac system strength and ac impedance characteristics. In: *10th IET Int. Conf. on AC-DC Power Transmission*. 2012: 1-6.
8. Jing Zhang, et al. Steady-State Operating Range of MMC-HVDC. *Electric Power Construction* 36.3(2015):1-6.
9. Yanan Li, Weiyong Jiang, Shifeng Yu, et al. System Design of Zhoushan Multi-terminal VSC-HVDC Transmission Project. *High Voltage Eng ineering*, 2014, 40(8):2490-2496.
10. Ying Liu, Linlin Wu, Man Xu, et al. Study on Steady-state Operation Region of VSC-HVDC Converter Station Connecting New Energy Cluster by Isolated Network. *North China Electric Power*, 2017(06):8-13

Realization of A Knowledge-based Intelligent System for Power Dispatching Plan Management

Qi Liu[1], *Yaoyi* Xu[2,*], *Peng* Zhan[1], *Xin* Li[1] and *Qike* Wu[1]

[1] Power Dispatching and Controlling Centre, Guangzhou Power Supply Co., Ltd., Guangzhou 510620, China
[2] School of Electrical Engineering, South China University of technology, Guangzhou 510640, China

Abstract. With the expanding of power grid scale in Chinese metropolis, the task intensity of power dispatchers increases rapidly in regulation of the power system operation structure and states to deal with everyday scheduled maintenance. In this paper, we propose a knowledge-based intelligent system developed to deal with daily management of the power dispatching plans. The system will analyse all the operation state changing tasks arranged for the next day and group the plans according to their association. It will automatically check the security of each power dispatching plan and generate the corresponding dispatching-order tickets. The proposed system builds up power grid ontology knowledge and first-order logic rules and integrates techniques of knowledge reasoning, natural language understanding and network topology analysis. Application shows that it can effectively realize the day-ahead power dispatching plan management (PDPM) instead of the human dispatchers.

1 Introduction

With the arrival of the new wave of artificial intelligence (AI), the construction of intelligent system in power system is in sustained growth [1-6]. Power dispatching and controlling is one of the hot areas for AI application. One of the challenges is how to build up the knowledge system about power dispatching in the machine, so that the computer can carry out complicated power dispatching tasks with high accuracy and efficiency instead of human dispatchers.

At present, most of the developed systems can only deal with some modules of the task, such as the inference of the dispatching-order tickets [7] and the trend data generating for pre-dispatching [8]. Although the papers [9-11] propose the ideas of intelligent robot or new paradigm for power dispatching and controlling, they have no further discussion on their designs and implementation.

The power supply system of a city is usually in looped network structure, but operated in open-loop mode to control the short-circuit current. Therefore, the operation mode will be frequently changed along with the outage or maintenance of the lines or transformers. For fast developing metropolis in China, most of the daily workload in power dispatching center comes from the operation modes switch process, which is composed of complicated cross-checking series to ensure the safety of the system. Usually, staffs of the maintenance department submit the equipment maintenance application through the OMS (Outage Management System). Then, the technicians responsible for operation mode schedule and protection setting provide their opinions on the OMS too. Finally, on the day before the execution of the power dispatching plans, the dispatcher needs to check the plans and evaluate their feasibility according to the load-forecast results and the latest system operation states. For the plans that pass through the check, the dispatcher will formulate corresponding dispatching-order tickets for the next-day execution of the plan. The dispatching-order ticket is a description of the specific operation sequences for the plan. During the process, the dispatcher may also exchange opinions with the applicants and the plan-makers via OMS or telephone to inquire or confirm the relevant issues.

In this paper, we present an intelligent computer system (IS) for the power dispatching plan management (PDPM), which can take over the task of dispatcher to realize fully automated closed-loop processing of the dispatching plans. The framework and main modules of the IS are introduced in section 2. Designing of the knowledge system is presented in section 3. A case test results as well as the discussions are given in section 4 of the paper.

2 Framework and main module designs of the IS

2.1 System physical structure

The physical structure of the IS for PDPM is shown in Figure 1.

The task controlling center is in charge of the workflow optimization so that the system can deal with piles of plans in parallel processing. In the task controlling center, the message management module interacts with the OMS and EMS (energy management system) through message

* Corresponding author: xu_yaoyi@foxmail.com

queue. The task management module controls the items and sequence of tasks in the task lists.

The inference engine is activated by different functional modules for knowledge reasoning. For example, the dispatching-order ticket is generated by the rule inference according to the operation mode schedule.

Database stores the network structure, operation states, apparatus parameters and load data etc. In addition, the knowledge system is also stored in the relational data base.

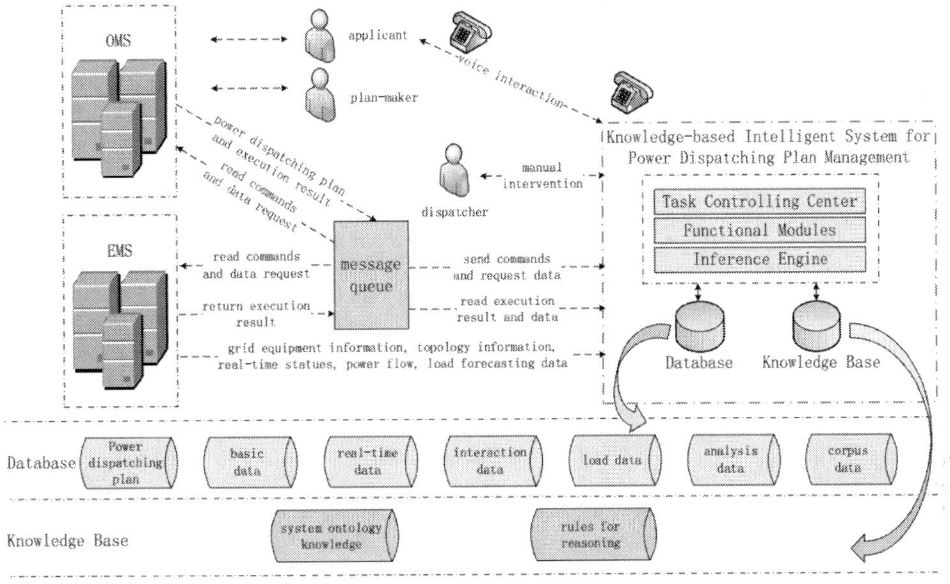

Fig 1. System physical structure.

2.2 Module functions and technologies

The functional modules of the IS and the corresponding technologies adopted are shown in Figure 2. We design an integrated natural language understanding (NLU) package to translate the operation mode schedules written in Chinese into computer accessible data forms, where the detail techniques are not introduced in this paper. According to the task information obtained from NLU, the facts about the related power equipment are indexed and transformed into a knowledge expression format that is suitable for computer reasoning.

The PDPM process procedures are decomposed into following functional modules: the associated plan grouping, the security check and dispatching-order ticket generating.

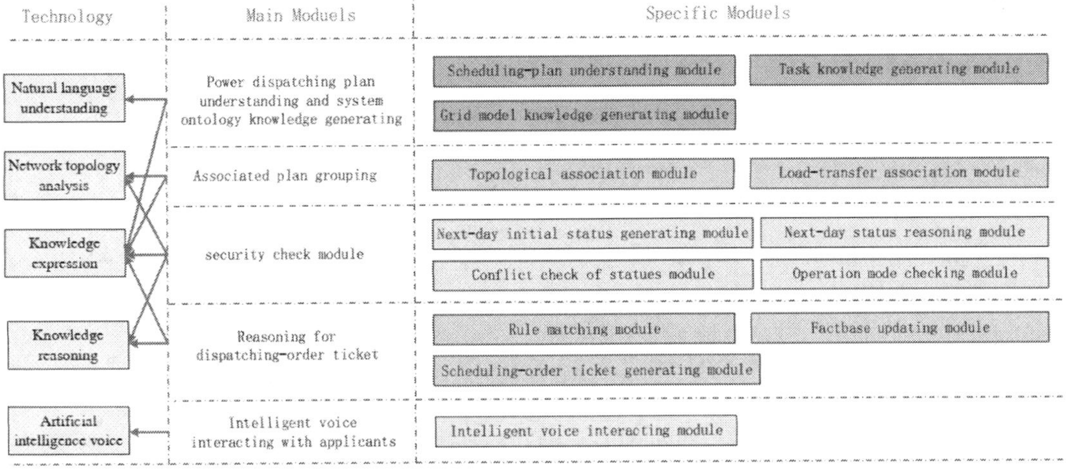

Fig 2. Functional modules corresponding implementation technologies of the system.

The associated plan grouping module (APG) works to identify the correlated plans. Different types of correlation are considered. The first one is the topological association, which means that different plans may have overlapped operating targets with conflicts, such as breakers. The load-transfer association is another factor, for instance, the load is transferred to the same transmission line or transformer in two different power dispatching plans. The checking of association covers equipment in different voltage layers from 220kV to 10kV, while the grid range of interest are obtained according to the plan relevant equipment and the topological information.

The security check module is carried out for each group of plans. At first, the breaker-status time sequences in the next day is generated according to the current operation states and the operation mode schedules. The rule-based reasoning is applied to infer the pre-operation and post-operation breakers and disconnectors statues. And the system will check if the statues of plans in each group have conflicts. Then the operation statues in the next day of each group are divided into 96 time sections (one interval for 15 minutes) with different load levels and operation network topologies. The load flow and load ratios of lines and transformers are calculated to verify the operation limits.

If the plans pass the check, the approved equipment target states will be added to the factbase and the dispatching-order ticket reasoning module will be activated. In this module, the inference engine applies the rule matching algorithm to generate the operation sequences according to the current facts. The designs of the fact and rule expression are introduced in the next section.

3 Grid knowledge expression based on the first-order logic

In the proposed IS, we adopt the first-order logic language for knowledge expression as shown in Figure 3. First-order logic can define the relationship for part objects or the whole domain which reduces the number of rules and simplifies the reason process. Here, all the knowledge about the power grid are divided into two large categories, the system ontology knowledge and the rules for reasoning.

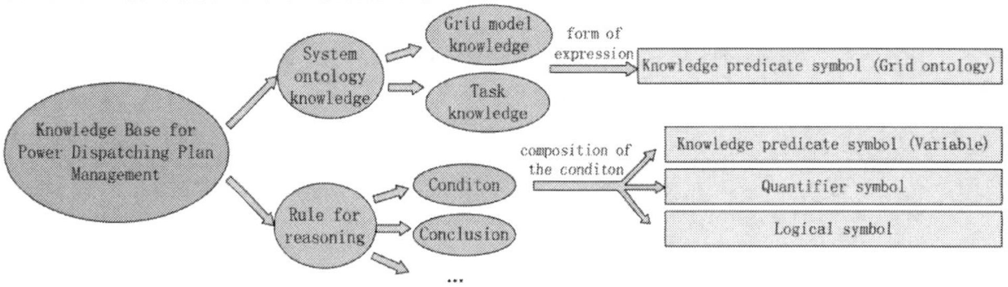

Fig 3. The division and composition of the system knowledge base.

The system ontology knowledge describes the grid equipment related information, whose expression form is the knowledge predicate symbol with parameters. The grid ontology is the real physical equipment in the grid, including power equipment, controlled devices and management departments etc. The knowledge predicate symbols are defined to represent the attributes, relationships and operation tasks of the power system. For example, the predicate symbol 'Transformer (A)' means that A is a transformer.

The rule for reasoning is composed of premise and conclusion. The premise is consisted of the knowledge predicate symbols, variables, quantifiers and logical symbols in first-order logic form, which shows strong description ability and versatility.

3.1 The establishment and expression of system ontology knowledge

We design a hierarchical classification framework for the system ontology knowledge. The ontology knowledge is further divided into two categories. One is the task knowledge obtained from the content of the power dispatching plan through the NLU module, which describes the power outage equipment and operation arrangements. The other is the grid model knowledge, derived from the database.

3.1.1 Grid model knowledge

The grid model knowledge includes the static conceptual knowledge, the dynamic conceptual knowledge and the object relation knowledge. The framework is shown in Figure 4, where the knowledge predicates uses the red fonts.

Static conceptual knowledge represents relatively fixed knowledge about the grid, such as the type of equipment, the type of work unit, the mode of bus connection and the target state of the relevant equipment, etc. The target state of equipment describes the state of equipment during the maintenance task.

The dynamic conceptual knowledge describes the dynamic information of the grid ontologies, including the real-time statues of the equipment, the power flow data, etc. They are updated through a data interface with the EMS.

Object relation knowledge represents the topological connection and affiliation of different elements. The affiliation defines the substation and voltage levels of equipment. The switchable equipment includes breaker, disconnector and the grounding switch.

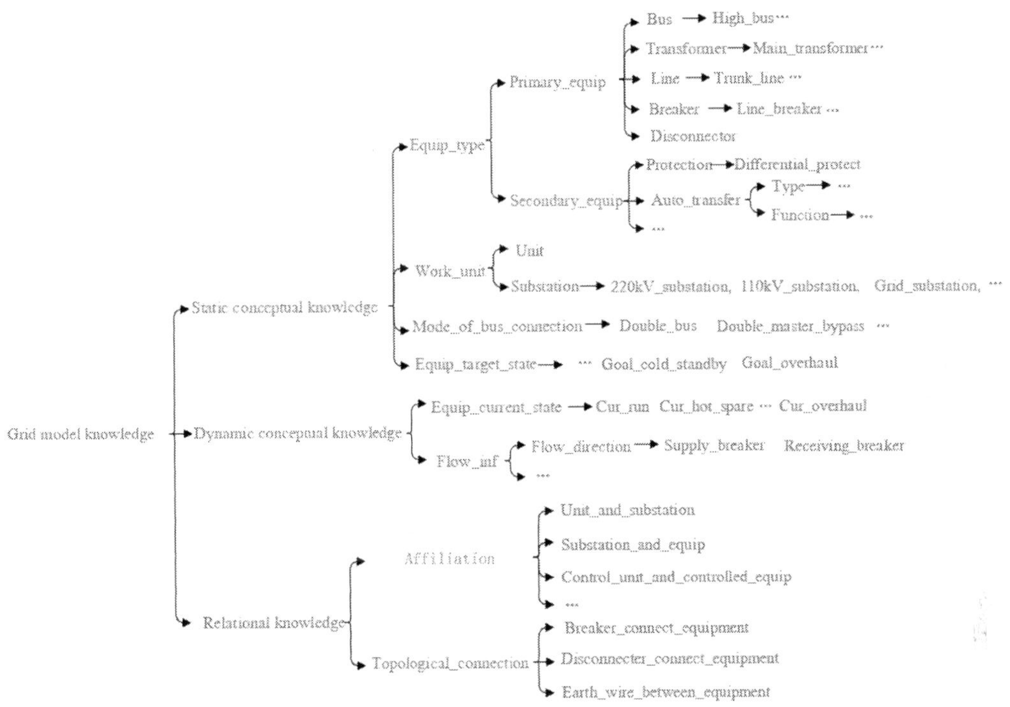

Fig 4. The organization and expression framework of Grid Model Knowledge.

3.1.2 Task knowledge

As shown in Figure 5, Task knowledge includes power outage equipment, state confirming tasks and operation tasks which describe different aspects of the plan. The state confirming tasks check the state of equipment in the kernel stage of operation, which are seldom reflected in the dispatching-order tickets. The operation tasks include load transfer, change of bus operation mode, etc. It describes the changes of the power system operation states or controlled equipment.

The hierarchical structure of system ontology knowledge proposed in this paper organizes the knowledge about the scheduled plan contents, grid topology, power flow information and equipment statues etc. into forms and groups that is easy for analysis and computer reasoning. The knowledge predicate symbols reflect concepts. For example, 'Transformer (B) 'and 'cur_run (B)' mean that the B is a transformer in operation.

In practical applications, the task knowledge and grid model knowledge related to each plan form the factbase of the plan, which is the source of the knowledge reasoning.

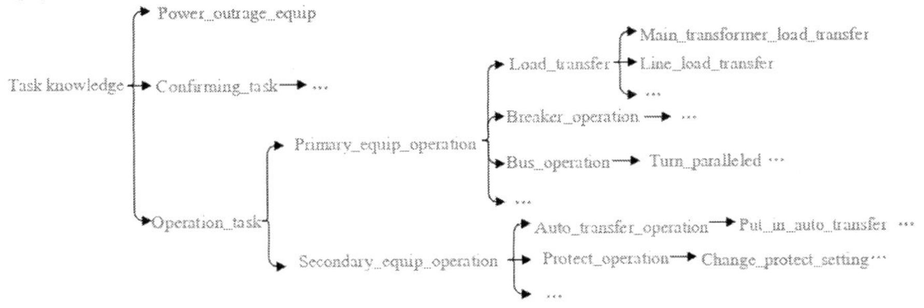

Fig 5. The organization and expression framework of Task Knowledge.

3.2 Design of rules for reasoning

The rule for reasoning adopts an enhanced IF-THEN expression as follows:

If: C_d, Then: (C_n, O_p, S_c), For: R_e, Priority: P

C_d is the rule premise, which is composed of several sub-conditions C_{di} and logic symbols such as '&' (and), '|' (or), '!' (not), '->'(contain), '='(equivalence), '∀'(full-term quantifier), '∃'(existence quantifier) and so on. The sub-condition $C_{di}(X_i,...)$ is composed of the knowledge predicate symbol C_{di} with variable X_i.

The rule conclusion consists of three parts: $C_n(X_i,...)$, $O_p(X_i,...)$ and $S_c(X_i,...)$.

$C_n(X_i,...)$ is usually an operation task and $(X_i,...)$ describes the operation targets. $S_c(X_i,...)$ defines the influence of the operation task on the state of the system. A general expression of $S_c(X_i,...)$ is 'State_A$(X_i,...)\rightarrow$State_B$(X_i,...)$'. $O_p(X_i,...)$ and R_e are in the form of Chinese natural language combined with variables. $O_p(X_i,...)$ is specially designed for the inference of the dispatching-order ticket, which gives the template to describe the operation followed the expression standards. R_e is designed for the explanation of the rule inference process so that the users can review and check the working mechanism of computer.

P means the priority of rules. In order to improve the reasoning efficiency, the inference engine will prefer to activate the rule with high priority value. There is an example of the rule, which is work for the dispatching-order ticket reasoning in table 1, where the Chinese expression components, i.e. $O_p(X_i,...)$ and R_e, are omitted.

Table 1. An example of rules for ticket reasoning.

C_d	C_n	S_c	P
∃x.(trunk_line(x)&current_run(x)&goal_overhaul(x)&∃y.(transformer(y)&current_run(y)&∃n.(self_substation(n)&∃m.(breaker(m)&transformer_belong(x,n,y,m)&∃z.(transformer(z)&transformer_load_transfer(y,z))))))	execute_transformer_transfer_load(y,z,n,m)	current_run(y)→ current_cold_standby(y), current_run(m)→ current_cold_standby(m)	4

It can be seen from the above table that C_d involves a large number of variables X_i, which work together to describe complex grid structure, operation mode and equipment relationship, etc. It makes the expression of rules completed, flexible and versatile, which can meet the application requirements of various operation tasks and complex operation modes.

In reason process, the inference engine matches the facts in the factbase with the premise of rules. The system builds in two categories of rules: rules for status reasoning and rules for dispatching-order ticket reasoning.

4 Case analysis

In this paper, we use an example to illustrate the functions of the system, where the structure of the corresponding sub-system is shown in Figure 6. The 110kV Line II supplies the 110kV transformer #1 in Substation V and the transformer #1 in Substation W. The 110kV Line IV supplies transformer #2 in Substation W. The 110kV bus bar sections of Substation W are supplied by other line separately.

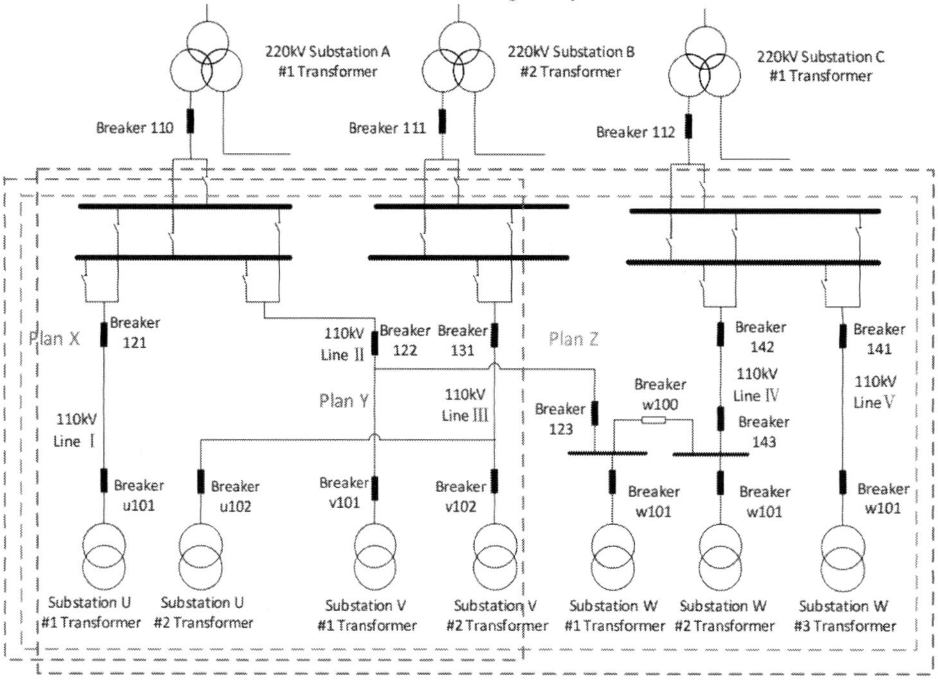

Fig 6. Topology of the case.

The IS received three plans for the next day on the day ahead (September 1st), as shown in Table 2. Technician for the operation mode schedule has submitted their opinion for operation mode changes during the equipment maintenance, which is written in Chinese and has been translated into English in column 3 of table 2. Unless otherwise specified, the operation mode will restore to the same as that before the maintenance.

Table 2. Examples of power dispatching plan.

Plan Number	Power outage equipment	operation mode schedules	Power outage time	Restoration time
X	110kV Line I and Substation U #1 Transformer	Transfer the load of Substation U #1 Transformer to #2 transformer	September 2nd 8:00	September 2nd 18:00
Y	110kV Line I and Substation V #1 Transformer	Transfer the load of Substation V #1 Transformer to #2 Transformer. Turn 110kV bus in Substation V into parallel operation	September 2nd 9:00	September 2nd 20:00
Z	Substation W #3 Transformer	Transfer the load of Substation W #3 Transformer to #2 Transformer	September 2nd 11:00	September 2nd 16:00

The system performs natural language understanding on each plan and then generates the task knowledge. After determining the range of interest in each plan according to the operation mode schedules (as shown by the dotted line in Figure 6), the IS generates the knowledge of the grid model. The task knowledge and grid model knowledge of each plan form its factbase.

Then the three plans are classified into one group since their interest ranges are overlapped. Plan X and Y are both related to 220kV Substation A and 220kV Substation B, and plan Y and Z are also both related to the 220kV Substation C. The grouping results from the associated power dispatching plan module are as following table 3.

Table 3. Grouping results of associated power dispatching plans.

Association group
Power dispatching plan X, Y, Z

After the table of next-day initial statues of breakers and disconnectors is formed, the reasoning for next-day statues is performed for each plan in the group. The system matches the facts in the factbase with the status reasoning rules. Some of the results are shown in Table 4, where the disconnectors statues are omitted for simplification and 0/1 refers to the breaker states of opened/closed.

Table 4. The breaker statues of each plan.

Plan number	Pre-operation breaker statues	Breaker operation statues	
X	121-1, u101-1, u102-1	8:00-18:00	121-0, u101-0, u102-1
		18:00-24:00	121-1, u101-1, u102-1
Y	122-1, v101-1, v102-1, 123-1, w100-0, ...	9:00-20:00	122-0, v101-0, v102-1, ...
		20:00-24:00	122-1, v101-1, v102-1, ...
Z	141-1, w103-1, ...	11:00-24:00	141-0, w103-0, ...

According to the inferred breaker statues, the system performs the check of breaker statues in each power dispatching plan:

(1) The system checks the pre-operation statues of each plan with the next-day initial statues table. If the statues are consistent, it will be proved that there is no conflict between the grid operation mode schedules of each power dispatching plan with the grid statues.

(2) The system performs conflict detection on the breaker operation statues of multiple power dispatching plans in the association group. Since there is no operation status of the same breaker in power dispatching plans in the group, it is confirmed that the operation mode contradiction does not exist among the power dispatching plans. Therefore, the correctness of each plan's operation mode is verified.

Then the next day's all-day breaker-statues in the group is divided by the interval of 15 minutes, as shown in the following table 5.

Table 5. Next day's all-day breaker-statues of the association group.

Time	Breaker statues
8:00	121-0, u101-0, u102-1, 122-1, v101-1, v102-1, ...
...	...
9:00	121-0, u101-0, u102-1, 122-0, v101-0, v102-1, ...
...	...
18:00	121-1, u101-1, u102-1, 122-0, v101-0, v102-1, ...
...	...

All 220kV grid partitions corresponding to plan X, Y and Z are taken as the topological range to be calculated. The table of the next-day initial statues of breakers and disconnectors is read as the initial statues in the calculation range. Calculation and judgment are performed according to the breaker statues of each state section and the load-forecast data of the equipment that is called in real time. The unqualified results in the calculation results are as table 6.

Table 6. Overload results of the association group.

Overload Time	Overload Equipment	Overload Numerical Value	Load of Related equipment	Related power dispatching plans
12:15	110kV Line III	0.56kA	Substation U #2 transformer：2.12kA； Substation V #2 transformer：2.45kA	X, Y

Based on above inference, the system sends back plan X and Y to the department for operation mode arrangement with suggestion that arrangement for the 10kV bus transfer should be added. Only after revision, these two plans can be re-approved.

Since there is no overload on the Main transformer #2 in Substation W, the operation mode check of plan Z will pass. After obtaining the equipment target states of plan Z, the IS performs the reasoning of the dispatching-order ticket. For the outage of the #3 main transformer in Substation W, the result is shown in the following table 7.

Table 7. Dispatching-order ticket of plan Z.

Sequence	Operation sequence	Substation	Dispatching order
1	1	Substation W	Transfer the load of #3 Transformer to #2 Transformer
2	1	Substation C	Confirm 110kV Line V has the dispatching operation conditions
3	2	Substation C	Disconnect the Breaker 141 on the 110kV Line V by remote controlling
4	3	Substation C	Turn the state of Breaker 141 on the 110kV Line V from hot spare to cold standby
5	4	Substation C	Turn the state of 110kV Line V from cold standby to overhaul

The above analysis example verifies the function and effectiveness of the dispatching plan intelligent processing system proposed in this paper. The system is now on the field test and the total process for day-ahead management is fast and reliable, which greatly improves the efficiency of the work.

5 Conclusion

This paper proposes a knowledge based intelligent system for day-ahead power dispatching plan processing and decision making, which can replace the human dispatcher in the corresponding work. The paper focuses on the design of knowledge expression and the module construction of the system. An example is given to verify the effectiveness of the proposed scheme.

The research work is supported by the Guangzhou Power Supply Co. of China under Grant no. GZHKJXM20170059.

References

1 C. Darab, R. Tarnovan, A. Turcu and C. Martineac.Artificial Intelligence Techniques for Fault Location and Detection in Distributed Generation Power Systems. In: 2019 8th International Conference on Modern Power Systems (MPS), 2019: 1-4.

2 L. Wu, J. Gao, G. K. Venayagamoorthy and R. G. Harley. On Artificial Intelligence Approaches for Contingency Analysis in Power System Security Assessment. In:2018 IEEE Power & Energy Society General Meeting (PESGM), 2018:1-5.

3 C. Tong, Y. Cai, Z. Zhang, Q. Wang, Y. Gao, J. Li, et al.. Artificial intelligence-based lightning protection of smart grid distribution system. In:2017 International Symposium on Lightning Protection (XIV SIPDA), 2017: 279-286.

4 H. Xu, J. Lu and Z.Yang. CiteSpace Based Knowledge Mapping Research of Artificial Intelligence Technology in Power System. In:2018 International Conference on Power System Technology (POWERCON), 2018:4383-4389.

5 A. Purwar, D. Joshi and M. S. Dasgupta. Smart control of electric lamp using artificial intelligence based controller. In:2015 Annual IEEE India Conference (INDICON), 2015: 1-5.

6 E. O. Kontis, T. A. Papadopoulos, M. H. Syed, E. Guillo-Sansano, G. M. Burt and G. K. Papagiannis.Artificial-Intelligence Method for the Derivation of Generic Aggregated Dynamic Equivalent Models. IEEE Transactions on Power Systems, 2019; 34(4):2947-2956.

7 S. Chen. Dispatching operation sheet system of power network based on multi-agent system. In: Proceedings of 2011 International Conference on Electronic & Mechanical Engineering and Information Technology, 2011:3754-3757.

8 J. Liu, Y. Wang, M. Lu, P. Xu, F. Lil and J. Geng. A Practical Approach to Generating Trend Data of Power Grid Operation for Pre-dispatching System. In:

2018 2nd IEEE Conference on Energy Internet and Energy System Integration (EI2), 2018: 1-4.

9 SHAN Xin, LU Xiao, ZHAI Mingyu, et al.. Analysis of Key Technologies for Artificial Intelligence Applied to Power Grid Dispatch and Control. Automation of Electric Power Systems, 2019; 43(1):49-57.

10 ZHANG Xiaohua, FENG Changyou, WANG Yongming, et al.. Design Ideas of Robotic Dispatcher for Power Grid. Automation of Electric Power Systems, 2019; 43(13):1-8.

11 J. J. Zhang, F.Wang, Q Wang, D. Hao, X. Yang, D.W.Gao et al.. Parallel dispatch: a new paradigm of electrical power system dispatch, IEEE/CAA Journal of Automatica Sinica, 2018; 5(1):311-319.

Probabilistic evaluation of renewable energy curtailment under forward electricity market

Jiheng Jiang[1] *, *Ying* Qiao[1], *Zongxiang* Lu[1], *Liang* Ran[2], *Ming* Ma[2], *Jin* Li[3], *Feng* Wang[2]

[1]State Key Lab of Control and Simulation of Power Systems and Generation Equipments, Dept. of Electrical Engineering, Tsinghua University, Beijing 100084, China
[2]State Grid Gansu Provincial Electric Power Company, Lanzhou, 730000, China
[3]Key Laboratory of Wind Power Integration Operation and Control of Gansu Province, Lanzhou 730070, China

Abstract. Forward electricity market has emerged as a pivotal section for the electricity transaction to keep stakeholders away from price risk and electricity shortage. But the forward contract of conventional unit partially limits the unit output before operation, leading to the curtailed generation share of renewable energy source (RES). Modeling and assessing this impact is of great significance for system planning and market supervision. The central thesis of this paper is to find out the impact of forward market on RES curtailment. A probabilistic evaluation model for RES accommodation is proposed, taking the curtailment rate as a main evaluation index. We mainly research financial contract and physical contract, modeling the impact of them on thermal unit minimum load capacity and power load in evaluation model. The simulation is conducted in a simplified system, which reveals the change of RES curtailment with renege penalty, contract price and execution generation curve.

1 Introduction

The forward market is an important part of the power market, acting as a complement for the spot market. The forward market refers to the market in which forward contracts are traded, and transactions are settled on a certain date in the future. The content of forward contract is usually the stakeholders agree to buy or sell a certain quantity of electricity at a certain price in the future [1]. So forward market plays a vital role in avoiding the price risk of electricity and detecting the demand in advance and was established in many countries.

Forward market is a classic market form in Chinese power system. For example, annual electricity plan and load entity direct transaction are the typical forward transaction without full marketization. With electricity marketization in China, forward market has become a central issue. In some provinces, the contract between thermal unit and load entity increase the lower bound of conventional unit forced power, leading to reduced generation share of RES. A primary concern of forward market here is how to evaluate the effect of pre-contracted electricity on future RES curtailment in a relatively long term, such as a month or a season. This issue is of great significance within the field of future power market construction.

Previous research was mainly focused on the market situation and economic analysis of forward market. To date, several studies have investigated the transaction mode of forward market in power market reform starting at 2015 in China [2], [3]. Reference [4] builds a new dynamic model to evaluate the effect of forward market on cyclical price behavior in electricity markets. An equilibrium model for joint forward contracts and day-ahead markets are proposed in [5], and the complicated market behavior of producers and consumers are also modeled. Agent-based model is a fundamental path to model the price building, transaction and operation of the forward market, which is used in [6]. In this study, authors consider renewable energy situation as boundary conditions but not dependent variable. Similarly, research [7] considers renewable energy volatility as an independent variable and indicates RES may undermine the role of forward contract, decreasing incentive of power plants. Another hot topic is the coordination of unit commitment and forward contract, on which reference [8] and [9] propose multi-time-range model for daily operation and long-term trading plans. Reference [10] analyzes the forward contract of interruptible load and gives an optimal management method, including purchase process and dispatch rule. The effect of forward market are only the auxiliary work of the above papers. What is not yet clear is the constrain of thermal unit minimum generation share caused by forward market and impact of forward market on RES curtailment in a long time scale rather than sequential analysis.

This paper attempts to explore this issue by building a probabilistic model based on probabilistic production simulation (PPS) method to assess RES accommodation. The two types of forward market considered are financial contract and physical contract. This paper begins by the probabilistic evaluation model for RES curtailment in

* Corresponding author: jjh17@mails.tsinghua.edu.cn

section 2. It will then go on to the model of financial contract and physical contract in the perspective of probability, described in section 3. Section 4 presents the result of numerical experiment that shows the change of RES curtailment with renege penalty, contract price of financial contract and execution curve shape, contract electricity of physical contract.

2 Basic principle of evaluation model

2.1 Curtailment power evaluation model

The main step to assess the impact of forward contract on renewable energy generation is building the evaluation model of renewable energy accommodation. Based on PPS, we illustrate the basic principle of generation curtailment and accommodation evaluation from a probabilistic perspective. Assuming that the sequential power load curve could be collected from TSO and be relatively stable in probability characteristic, which makes it possible to predict the future load curve on the basis of historical curve multiplied by rate. According to sequential load curve, the equivalent load duration curve (ELDC) $F_L(x)$ is defined as

$$F_L(X)=\int_X^{P_{Lmax}} f_L(x)dx \tag{1}$$

where x is value of power load or capacity of generation. $P_{L\max}$ is the maximum load, MW. $f_L(x)$ is the probability distribution function (PDF) of power load, calculated from historical data.

Power balance in PPS is the process that power sources fill the area between ELDC and X-axis in a specific order, shown in Fig. 1. Physically, this area is the total energy needed, or the accommodation area for sources. In Fig. 1, power source A, B, D and E are arranged in different positions under ELDC, where the width of each position is the capacity of generation. Source A,B and E are single state units, which means there is only one forced outage rate (FOR) in whole capacity, while source D have three different FOR in the variable states. If ELDC is less than 1, the generation arranged in that positon may face with curtailing energy, represented by yellow part in Fig. 1. Curtailment energy E_{cur} depends on the yellow part and FOR of corresponding units, which can be stated as

$$E_{cur}=T\bullet\sum_I(1-q_i)\bullet\int_{x_i}^{x_i+C_i}(1-F_L(x))dx \tag{2}$$

where I is the state set of a unit, and T is the calculation time range, usually a month. q_i is the FOR of state i. C_i is the capacity of state i. x_i is the starting point in the X-axis of state i.

Correspondingly, accommodation energy of this unit is

$$E_{acc}=T\bullet\sum_I(1-q_i)\bullet\int_{x_i}^{x_i+C_i}F_L(x)dx \tag{3}$$

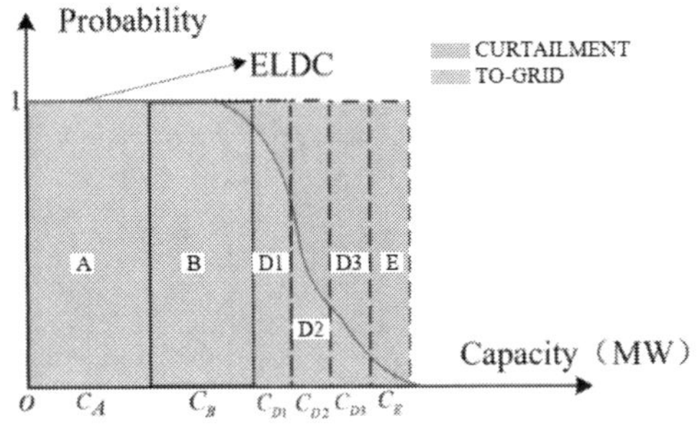

Fig. 1 Evaluation principle of generation curtailment and accommodation

It is obvious that the energy calculated by (2) and (3) are expectation value in a period. The process are independent from the sequential data, simplifying load prediction. As for unit types, totally controllable units, including conventional thermal unit, nuclear unit, are usually modeled as one-state or two-state unit which consists of forced capacity and regulation capacity. Curtailment energy is surplus generation capability but no fuel waste for them. So we usually focus on the curtailment energy of RES. Wind power and PV are usually types of multi-state, because they are constrained by natural resource. In that case, the capacity-probability pairs set of renewable energy can be calculated from PDF with discretization, shown in Fig. 2.

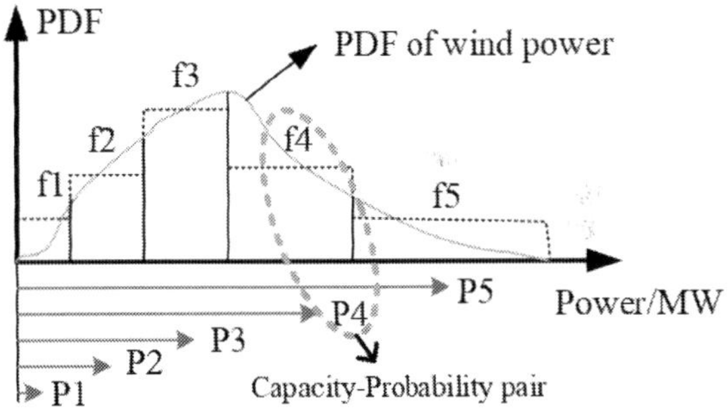

Fig. 2 Discretized capacity-probability pairs of wind power

2.2 Interaction and arrangement of power sources

With reference to the dispatching principle, the base load, especially part A and part B in Fig. 1, is supported by thermal units and nuclear units. And renewable energy is regarded as "middle power", supporting the shoulder load that is represented by part D in Fig. 1 for RES can't provide regulation service to track load. In PPS, the wind power will be filled after the minimum load capacity of the non-heating thermal power are involved. Sometimes, hydropower station with insufficient reservoir is arranged in the same position with wind power. However, this paper is concentrated on the fluctuant RES, mainly the wind power, without detailed research on hydropower. After that, the load not supported is peak load that is the main generation area for thermal unit regulation capacity.

The existing studies show that a large number of conventional units with too much minimum load will cause severe curtailment, which indicates the base load part, such as part A and part B in Fig. 1, occupies energy area of RES. The amount of power consumption shown in Part D of Fig. 1 will decrease as the thermal unit capacity increases. The closer to peak load the arrangement position is, the more possibility to curtail power the unit has. Therefore, the PPS method has a principle advantage in the evaluation of medium and long-term.

Describing the interaction of conventional power and RES, we still need to research how FOR influence the ELDC. If FOR of unit A is less than 1, it indicates this unit may black out randomly, adding power load to other units arranged behind A equivalently. So ELDC must be modified after a unit is arranged to support load. The modification formulation is

$$F^{(i)}(x)=\sum q_i F^{(i-1)}(x-C_i+C_E),(i=1,2,3,4\cdots) \qquad (4)$$

where $F^{(i)}(x)$ and $F^{(i-1)}(x)$ is the new ELDC and former one. C_i is the available capacity of state i. C_E is the rate capacity of arranged unit.

The ELDC will become more flat and higher, increasing generation area for next unit, for there is almost no unit with zero-FOR. In particular, this tendency is the most distinct after modified by RES.

3 Probabilistic model of forward market

3.1 Financial contract

For the purpose of financial contract in forward market, the design is only for avoiding the risk of energy price fluctuation rather than energy production uncertainty. Existing research has an assumption that the financial contract has no impact on real-time dispatch result. In that case, RES generation area will not be reduced by these virtual contract. In other word, financial contract may be decoupled from physical production. The previous model sounds reasonable but not practical. The unbalance penalty after contract period is becoming higher and price gap between contract and real time caused by congestion is highly fluctuant. We can conclude that high accounting penalty could form invisible constraints on physical execution.

To model this constrain effect, we assume the penalty would increase the minimum load capacity of a thermal unit bound with a load entity by forward contract. Part A in Fig. 1 may be broadened, if unit A is in a contract and the penalty is high. The modified minimum load capacity $C_{\min}$ is defined as follows:

$$C_{\min}=\begin{cases} C_{\min}^{ori}, & (\frac{Q_f}{T}\le C_{\min}^{ori}) \\ C_{\min}^{ori}+\eta\bullet(\frac{Q_f}{T}-C_{\min}^{ori}), & (\frac{Q_f}{T}> C_{\min}^{ori}) \end{cases} \qquad (5)$$

where $C_{\min}^{ori}$ is the physical minimum capacity of unit. Q_f is the contract energy in period T. η is the modification coefficient for $C_{\min}$. Then, η is proportional to penalty price and inversely proportional to contract price, which is stated as

$$\eta=\frac{2}{1+e^{-(\frac{m_p}{m_e}\bullet\frac{m_{es}}{m_{ps}})}}-1 \qquad (6)$$

where m_p is the average penalty price for unexecuted energy per MWh. And m_e is the average price of contract energy per MWh. m_{ps}, m_{es} that are pre-determined before PPS are the reference value of m_p and m_e.

In addition, the model proposed above links market price to system physical operation and implies that the minimum load capacity of thermal unit may increase at any time. That's why the minimum load capacity part under ELDC is still a rectangle.

3.2 Physical contract

The section above introduces a power source modification model for financial contract. As for physical contract, this section will propose an ELDC modification model. Physical contract between power plant and load entity contains detailed contract time, power and price, which can be seen as a deterministic and fixed daily load curve. The following analysis uses unit A as an example to illustrate the modeling method.

The time when physical contract of unit A is operated in a day is $[0,t_s]\cup[t_e,24]$. We just research units only with a physical contract but no financial contract. So the intraday load sequence $L(t)$ is separated into two series:

$$L_1(t)=\begin{cases} 0, & t\in[0,t_s]\cup[t_e,24] \\ L(t), & t\in[t_s,t_e] \end{cases}$$
$$L_2(t)=\begin{cases} L(t), & t\in[0,t_s]\cup[t_e,24] \\ 0, & t\in[t_s,t_e] \end{cases} \qquad (7)$$

$F_{L1}(x), F_{L2}(x)$ are the ELDC of $L_1(t), L_2(t)$. And $f_{L1}(x), f_{L2}(x)$ are PDF (refer to formula(1)). Physical contract works on $L_2(t)$ and reduces the accommodation area for RES. So $F_{L2}(x)$ and $f_{L2}(x)$ must be modified before the calculation of PPS. We define $L_2^{new}(t)$ is

$$L_2^{new}(t)=L_2(t)-L_p(t) \qquad (8)$$

where $L_p(t)$ is the equivalent power curve of physical contract applied in PPS. It eliminates the minimum capacity because only the contract load part overtop the minimum load capacity will decrease the area for RES.

$$L_p(t)=\begin{cases} L_p^{ori}(t)\text{-}C_{\min}^{ori}, & L_p^{ori}(t)\ge C_{\min}^{ori} \\ 0, & L_p(t)< C_{\min}^{ori} \end{cases} \qquad (9)$$

where $L_p^{ori}(t)$ is actual contract load curve.

A major assumption is the power load is regular in long term, which means the power level and PDF of intraday load in $[0,t_s]\cup[t_e,24]$ is almost the same in a month. And the contract power is totally fixed in the

 https://doi.org/10.1051/e3sconf/202018202005

same time everyday. The (L_2, L_P) is joint random variable of load and contract power. And the PDF of this variable is defined as $f_L(l_2, l_p)$. The PDF of $L_2^{new}(t)$ is defined as $f_L(l_2^{new})$ calculated by following convolution formula.

$$f_L(l_2^{new}) = \int[f_{L.l_2}(l_2^{new}+l_p)\cdot f_{L.l_p}(l_p)]\,d\,l_p \qquad (10)$$

where $f_{L.l_2}(.)$ is the edge distribution of l_2 for PDF and $f_{L.l_p}(.)$ the edge distribution of l_p.

Then, $F_{L2}(x)$ is corrected to $F_{L2}^{new}(x)$ which is the integral of $f_L(l_2^{new})$. And the modified ELDC $F_L^{new}(x) = F_{L1}(x) + F_{L2}(x)$ is generated. In practice, $f_L(l_2, l_p)$ can be calculated through statistics. Two edge distribution functions are easy to get from $f_L(l_2, l_p)$. Even though the load data is discrete rather than continuous, the principle of integral is still applicative.

3.3 Process of forward contract effect evaluation

The process of PPS for evaluation calculation is stated as following.

Step 1. Generate the ELDC based on the historical data of generation load in the region.

Step 2. For all conventional thermal units, two sections of every unit are separated in accordance with the minimum power and adjustable power.

Step 3. Calculate the equivalent minimum load capacity of thermal units that are in financial contract through (5). Arrange the minimum power part of the conventional thermal units to support the basic load.

Step 4. Modify the ELDC according to the fixed load curve in physical contract through formula (7)-(10).

Step 5. Arrange wind power into production.

Step 6. Calculate the abandoned wind power energy E_{wc} through formula (2).

Step 7. Filled the rest area with the adjustable power output of conventional thermal units.

4 Simulation and results

4.1 Data set

A simplified simulation system, consisting of three thermal power units, an equivalent centralized wind farm and several load entities, was used to analyze the effect of forward market. The total capacity of power source is 2050MW. Table 1 shows details of power source and system generation load.

Table 1 Parameters of power source and load in system

Thermal			Wind power			Load		
Unit No.	Capacity(MW)	Min load (MW)	Max(MW)	Min(MW)	Average(MW)	Max(MW)	Min(MW)	Average(MW)
1	600	240	596.8	122.7	481.2	1203.8	737.6	985.7
2	300	120						
3	300	120	Rated capacity: 850MW			Total energy: 709.7GWh		

Sequential load curve and wind power curve of 15-minute time precision are proportionately modified from the data of typical province power system in northwestern China. The simulation is conducted in one month. Fig. 3 (a) shows the typical curve of load and wind power supported by thermal unit minimum load capacity in first 5 days. Fig. 3 (b) shows the PDF of wind power. The result reveals that there has been a slight RES curtailment with all thermal units on. The curtailment energy of whole month is 22852.6MWh and curtailment rate is 6.05%.

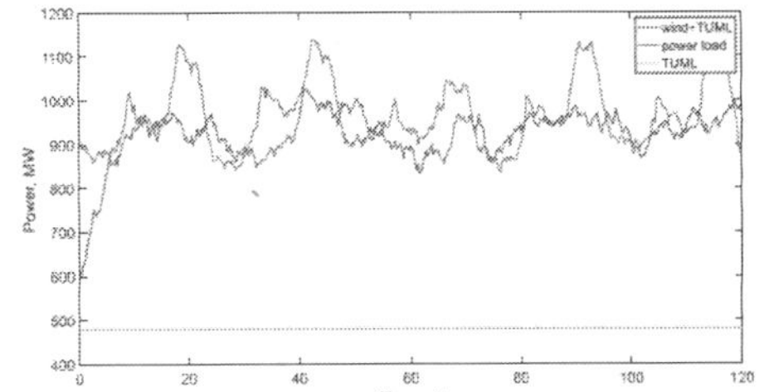

(a) Typical curve of load and wind power (TUML-Thermal unit minimum load)

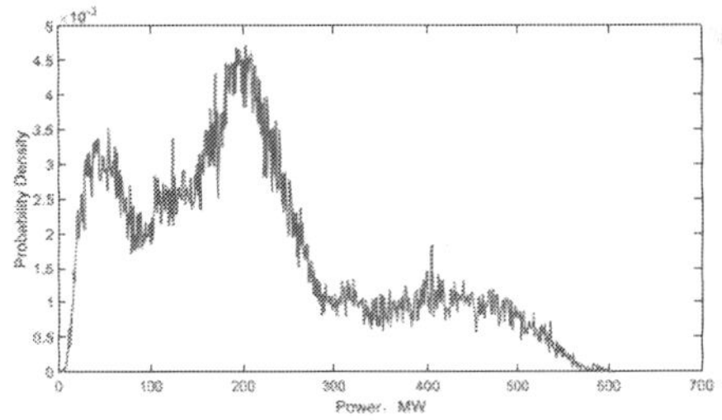

(b)PDF of wind power

Fig. 3 Sequential curve and statistics of power source

4.2 Effect of financial contract

In this numerical trial, we assume only thermal unit No.2 has a financial contract with other load entities but no physical contract. Fig. 4 shows the value change of η with variation of penalty and contract price according to formula(6). The reference value of contract price and renege penalty are 1 and 10 yuan/KWh, respectively. The figure can show that the increase of penalty and decrease of contract price will make coefficient climb up.

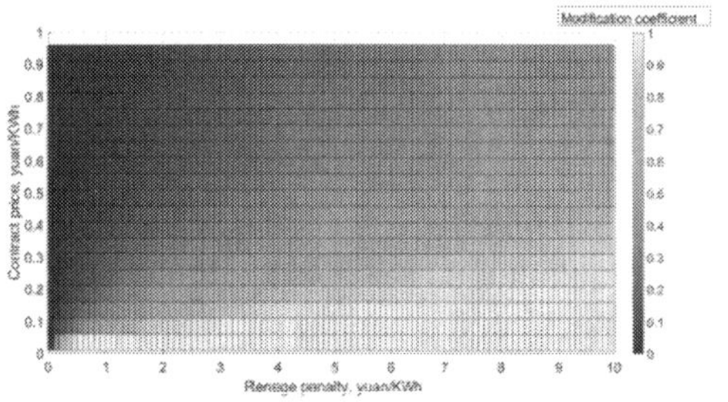

Fig. 4 Trends of modification coefficient η with renege penalty and contract price.

Fig. 5 shows the trends of RES curtailment rate when the penalty and contract price are variable. The reference value of contract price and renege penalty here are 1 and 6 yuan/KWh, respectively. Total contract energy of unit 2 is 108GWh in the month. The surface is similar to curved fan-shape, where positive correlation was found between curtailment and penalty price with a negative one between curtailment and contract price.

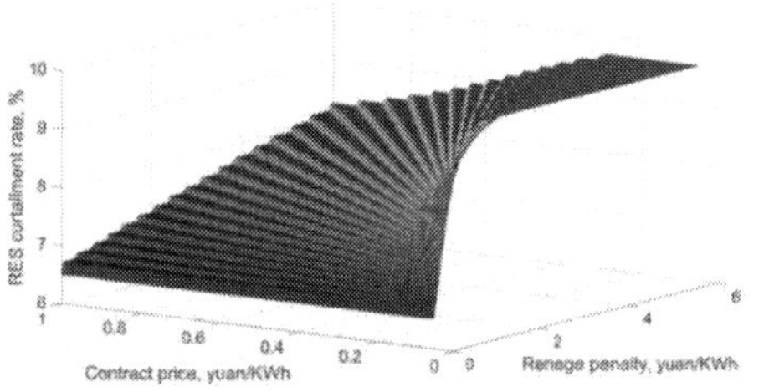

Fig. 5 Change of RES curtailment rate with renege penalty and contract price (contract energy is fixed in 108GWh)

The results of relationship between curtailment rate and penalty price under different contract energy is shown in Fig. 6 when the contract price is kept on 0.4 yuan/ KWh. When the contract energy is 86.4GWh, meaning the minimum load capacity of unit 2 will cover the contract demand, there is no effect on original curtailment. The extreme scenario will happen when contract energy is 216GWh, maximum generation energy in a month of unit 2, with penalty price of 6 yuan/KWh, leading to more than 30% RES curtailed.

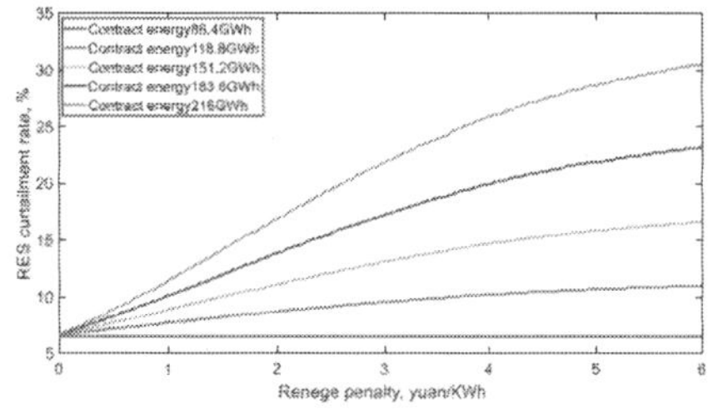

Fig. 6 Change of curtailment rate with penalty price under different contract energy

4.3 Effect of physical contract

The assumption in this case is only thermal unit 2 has a physical contract with a fixed execution load curve but no financial contract, which must be dispatched in operation. And the curve is the same in every day. We applied four typical shapes of contract curve in the case, displayed in Fig.7 (a). All the typical curves correspond to the same contract energy of 106GWh in one month. The curve of mode A has a similar tendency to system load with double peak-valley shape, while mode B is with reversed double peak-valley shape, mode C with fixed value and mode D with single peak-valley shape. The notable impact on the ELDC is found that mode A mainly decreases the rear tail of ELDC and mode B reduce the former tail, shown in Fig.7 (b).

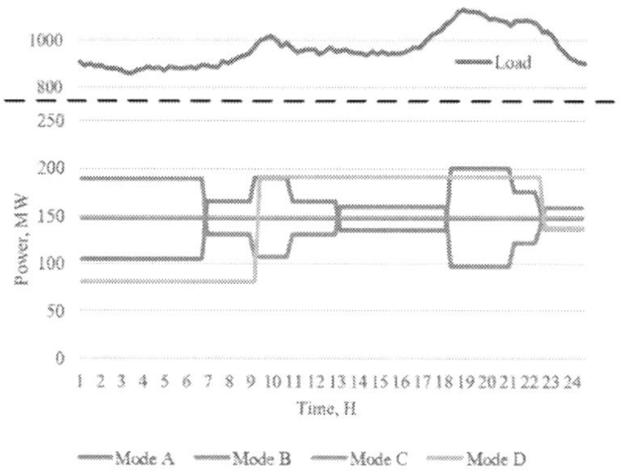

(a)Typical execution curve and system load

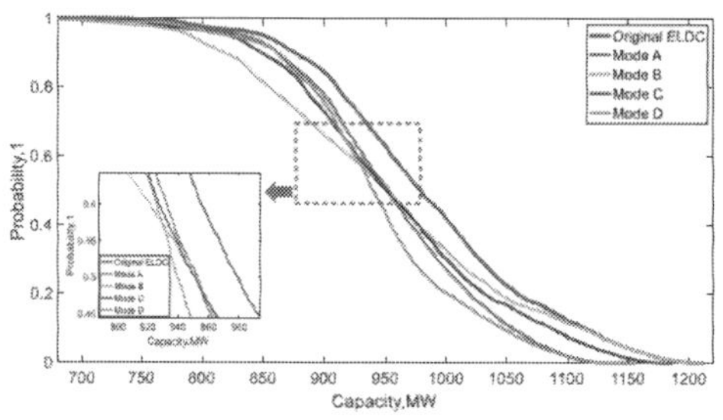

(b)The impact of physical contract on ELDC

Fig. 7 Typical execution curve of physical contract and impact on ELDC

The curtailment rate and energy with different modes are listed in Table 2. Compared to Fig. 5, we can see with the same total contract energy, physical contract has more severe impact. But the impact of mode A is the lightest and mode B the worst. It is apparent from this case that the curve shape impact curtailment a lot, and the shape in anti-load tendency is the worst choice.

Table 2 Curtailment rate and energy with four curve modes

	Oringinal	Mode A	Mode B	Mode C	Mode D
Curtailment rate/%	6.05	**8.26**	10.83	8.93	9.29
Curtailment energy/MWh	22852.6	**29014.4**	38043.7	31387.4	32645.1

https://doi.org/10.1051/e3sconf/202018202005

A further trial is conducted in a larger data range where the curves in Fig.7 (a) are chosen as the reference and proportionally change the contract electricity from 50% to 150%. The curtailment rate with four modes under different ratios is shown in Fig. 8. It can be found that the curtailment with mode A is the slightest when the ratio is between 90% and 150%. Meanwhile, the result approves mode B is the worst contract mode whatever the ratio is.

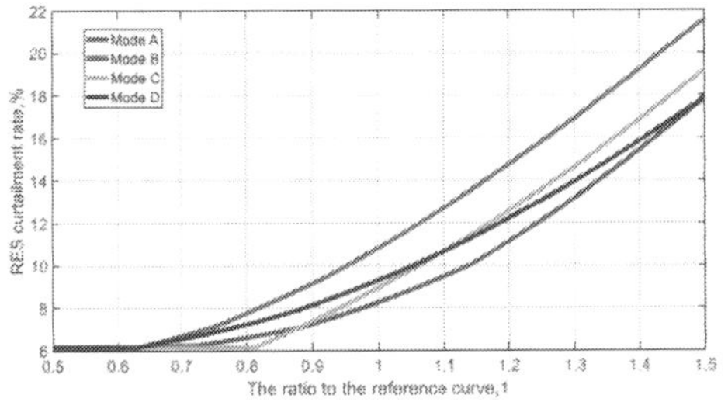

Fig. 8 RES curtailment rate with four types of curve in different ratio (compared to reference curve in Fig.7 (a))

5 Conclusion

This study set out to research the impact of forward market on system RES curtailment. The main types of forward market we studied are financial contract and physical contract where the stakeholders are thermal units and load entities. The evaluation model was built based on PPS, considering the probabilistic characteristic of forward contract, to calculate the RES curtailment rate in a month under different contract conditions. The experiments confirmed that forward contract of thermal unit will take up the generation share of RES. And physical contract has more impact than financial contract with the same contract electricity. The second major finding was that the impact of financial contract is sensitive to renege penalty and contract price, and physical contract sensitive to execution load curve. Taken together, these results suggest that the execution mode, price limitation and electricity decomposition curve of forward market must be considered carefully with the rapid increase of RES share. The study contributes to our understanding of relationship between forward market and RES accommodation. Further studies need to be carried out in order to research the influence of load entity price-response characteristics on the RES curtailment under forward market.

This work was supported by the State Grid Gansu Provincial Electric Power Company Key Project (SGGSKY00FJWT1900170).

References

1. K. Ruddell, A. Downward, A. Philpott. Market power and forward prices, *Economics Letters*, **166**: 6-9(2018).
2. S. Li, S. Zhang, P. Andrews-Speed, Using diverse market-based approaches to integrate renewable energy: Experiences from China, *Energy Policy*, **125**: 330-337(2019).
3. Xu R, Song Z, Tang Q, Yu Z., The cost and marketability of renewable energy after power market reform in China: A review, *Journal of Cleaner Production*, **204:** 409-424(2018).
4. K. C. Álvarez-Uribe, S. Arango-Aramburo, E. R. Larsen, Forward contracts in electricity markets and capacity investment: A simulation study, *Utilities Policy*, **54**: 1-10(2018).
5. M. Banaei, M. Oloomi Buygi, H. Raouf Sheybani, Supply function Nash equilibrium of joint day-ahead electricity markets and forward contracts, *International Journal of Electrical Power & Energy Systems*, **113**: 104-116(2019).
6. P. C. Bhagwat, A. Marcheselli, J. C. Richstein, E. J.L. Chappin, L. J. De Vries, An analysis of a forward capacity market with long-term contracts, *Energy Policy*, 2017, 111: 255-267.
7. Ritz, Robert A . How does renewables competition affect forward contracting in electricity markets. *Economics Letters*, **146**: 135-139(2016).
8. Miao S, Luo B, Shen J, et al. Short-term multi-objective hydro-thermal generation dispatch considering electricity market transition and mid-and long-term contract decomposition. *Power System Technology*, **42** (7) : 2221-2231(2018) .
9. Guo Y, Kuang H, Guo S, et al. Research on weekly unit commitment strategy harmonizing long term market trade planning and daily generation scheduling, *Power System Protection and Control*, **45** (22) : 78-82(2017) .
10. J. Wang, X. Wang and X. Ding, The Forward Contract Model of Interruptible Load in Power Market, *IEEE/PES Transmission & Distribution Conference & Exposition: Asia and Pacific*, Dalian, 1-5(2005).

https://doi.org/10.1051/e3sconf/202018202006

Knowledge representation and intelligent fault diagnosis technology for power grid faults

Xiang Li[1,*], *Xin* Min[2], *Zeng* Yang[1], *Hongyu* He[1], and *Jinfeng* Li[1]

[1] Power Dispatching and Controlling Centre, Guangzhou Power Supply Co., Ltd., Guangzhou 510620, China
[2] School of Electrical Engineering, South China University of technology, Guangzhou 510640, China

Abstract. Aiming at that the automatic fault diagnosis method is difficult to locate the fault causes under uncertain circumstances which include malfunctions of the equipment and wrong alarm messages, a knowledge model was proposed to describe the relationship, status and operation of the equipment. And based on the model, the action logic between equipment after accidents is expressed in the form of rules combined with predicate logic. The corresponding interpretation and checking results of the relative alarm messages are given by reasoning the accident chain under different fault hypothesis. And the optimal judgment result is obtained through the calculation of prior probability. The validity of the method is verified by a practical fault case.

1 Introduction

When the grid fails, the alarm messages would be uploaded to the dispatch centre through SCADA (Supervisory Control and Data Acquisition). The dispatchers judge the type of the fault based on the information, and then find the accident handling plan. However, the number of the messages is large, especially when the fault involved multiple substations. Therefore, some automatic fault diagnosis methods were proposed to achieve rapid screening of the alarm messages and determination of the fault causes which combine computer technology and artificial intelligence technology

The automatic fault diagnosis methods mainly include rule-based expert system [1-2], analytic model [3-4], Bayesian networks [5-6], Petri nets [7-8] and artificial neural networks [9-10]. The traditional expert system method expresses the relationship between protection and faulty equipment as production rules, and finds the cause of the fault by reasoning. This method is clear and interpretable. But the descriptive ability of the rules is finite and it usually gets combinatoric explosion under complex situation. And it is also difficult to carry out in the case of false alarms information and equipment malfunctions. Therefore, expert system method often cooperates with other methods such as information theory, which builds a channel model to describe the uncertainty in the information transfer process; The analytical method converts the fault facts and diagnostic rules into a mathematical model. And search the hypothesis that matches the fault feature best by solving the optimization problem. This method has a strict mathematical logic foundation and copes with the uncertain information in the fault well. However, the dimension of the model is usually too high and the relevant weights in the model are most based on subjective setting , so the efficiency and accuracy of the diagnosis cannot be guaranteed; Based on the artificial neural network method, the relevant protection and switching action information are taken as input, and the fault cause is taken as the output. The fault samples are used to train the diagnostic neural network, which avoids the explicit expression of the fault diagnosis knowledge. It has good tolerance for the reported error alarm information, but the interpretation of the result is poor and the training effect is difficult to guarantee under large-scale power grid. Petri net technology represents the topological relationship between devices and the logical action relationship between protection, circuit breakers and faulty devices in the form of directed graphs. It initializes the state of the petri network according to the fault information firstly and then the network state would be updated through continuous reasoning. Finally, the diagnosis result could be got under the steady state. The reasoning speed is fast, but it is less versatile and remodelling is required under different grid structures.

In summary, the current automatic diagnosis technology has two difficulties: firstly, the descriptive ability and maintainability of the knowledge model reflecting the fault judgment logic are difficult to combine. Most methods only consider the relationship between protection and primary equipment. The operation of the automatic device for power recovery after voltage loss is not taken into account. This makes it difficult to explain the complete process of accident evolution and affects the accuracy of judgment. Secondly, the information source of the accident judgment often has false or unreported message, and the

* Corresponding author: lix@guangzhou.csg.cn

 https://doi.org/10.1051/e3sconf/202018202006

related equipment also has the possibility of refusal and maloperation. These uncertain factors need to be considered in the automatic diagnosis of the accident. In view of the above two points, this paper firstly defined a fault diagnosis knowledge model that can fully describe the fault state, and then it achieves automatic diagnosis by reasoning the fault chain under the uncertainty based on the model. Finally, it found the most possible interpretation of the fault by calculating the probability of the chain.

2 The knowledge representation model of fault diagnosis

When a grid fault occurs, according to detecting the change in electrical quantity, the relevant primary and secondary equipment trigger the action response and achieve the isolation and recovery of the fault under a certain action logic. These data constitute the source information of the fault diagnosis. The fault diagnosis is to realize the mapping from the source information to the faulty device. So this requires some way to express the knowledge involved in the fault diagnosis, including the topology relationship between the devices, the relationship between the faulty equipment and the protection, the relationship between the protection and the action switch, the relationship between the protection and protection as well as the logic of the relevant automatic devices in the power grid. The entire accident evolution process can be fully described and the fault diagnosis can be realized only by expressing these logical relationships adequately and accurately. In this paper, based on the thought of the object-oriented, the abstract classification of object categories is performed according to the operational and functional characteristics of the grid equipment object in the situation of fault firstly. And then the ontology representation of fault diagnosis knowledge is formed by the logical association and attribute relationship between these object classes. Finally, through the instantiation of the grid object, the knowledge map of the fault diagnosis is gradually generated from the top to the bottom.

2.1 The hierarchical division of concepts in grid

The electrical quantity of the primary equipment changes suddenly when the power grid fails. At the same time, the secondary equipment detects the change of the operating parameters and triggers the action, controls the corresponding switch displacement to achieve the isolation of the faulty device and the recovery of the pressure-loss equipment. Based on the functional and operational characteristics of different device objects in the process, the relevant grid equipment is divided into categories in this paper, and the specific classification result are shown in figure 1.

As shown in figure 1, according to the different functions in the power grid, the equipment is divided into primary and secondary device, in which the primary device is classified into breaking device and operating device. Breaking device is the bearer for realizing the change of the operation mode of the power grid, and it is also the operation object of the fault treatment measures. According to the breaking capacity, it is divided into circuit breaker and disconnector here. Operating device is the main body of function implementation in the grid operation, but it does not have the ability to change the operating state. The operating parameters of these equipment are the characteristics of the grid operating conditions. They are the protection objects of the secondary equipment, which are mainly classified into bus, main transformer and line. In the secondary device, only the automatic device, protection and reclosure class are considered here according to the implemented functions. The automatic device is a control device for restoring the power supply of the voltage loss device after the fault, including bus automatic device and the line automatic device. Protection is a control device that isolates faulty equipment when a fault occurs. According to the different protection triggering principles, it is divided into distance protection, differential protection, overcurrent protection, etc. In addition, in order to cope with a large number of transient faults in the power grid, the lines often have reclosures. In order to be able to describe the relevant state, the reclosure class is defined here.

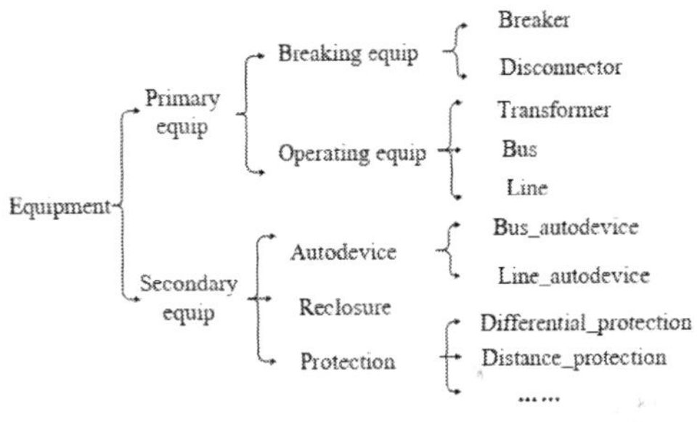

Fig. 1. Classification of electric network equipment.

2.2 The definitions of device properties and relationships

A large amount of operational data is generated in the real-time operation of the grid equipment. The information which the equipment attribute consisting of numerical values or strings reflects the operating status of the grid. On the other hand, there is a complex and close relationship between the grid equipment. After the fault occurs, the state change of the initial faulty equipment will trigger the action and change the attributes of all levels of equipment along the logical chain among the equipment, and finally change into the post-accident state. In order to express related information, based on the device concept categories classification as mentioned above, a constraint definition is defined on the attributes of each object concept and the relationship between the objects. And the representation format of the relationship between objects is defined as <object, attribute, attribute value, time>. Considering the time constraint relationship of the action logic between devices, a time label is added for each

 https://doi.org/10.1051/e3sconf/202018202006

attribute, which the time of the device action attribute is the action time, and the time label of the device status attribute is the measured time. The specific device attributes are defined as the following table 1.

Table 1. Object attribute definition

Attribute name	**Object class**	**Attribute value**
Switch_state	Breaking equip	Closed/Open
Operating equip state	Operating equip	Run/Maintenanc e
Secondaryequip _state	Secondary equip	Off/On
Functional_state	Equipment	Fault/Normal
Current value	Primary equip	number
Voltage value	Operating equip	number
Operation_1	Breaking equip	Trip/Closing/ No-action
Operation_2	Secondary equip	Action / No-action

The representation format of the relationship between objects is defined as <object, relationship, object>. The specific definition is as shown in figure 2, in which each type of object relationship is represented by a pointing line segment, from the relational subject to the relational object. For example, the object relation original source from automatic device (auto device class) points to function equip (operating equip class). The relationship subject that represents this relationship is constrained to the automatic device, and the relationship object is constrained to the operating device. The specific meaning is the relationship between the automatic device and its primary disconnected battery.

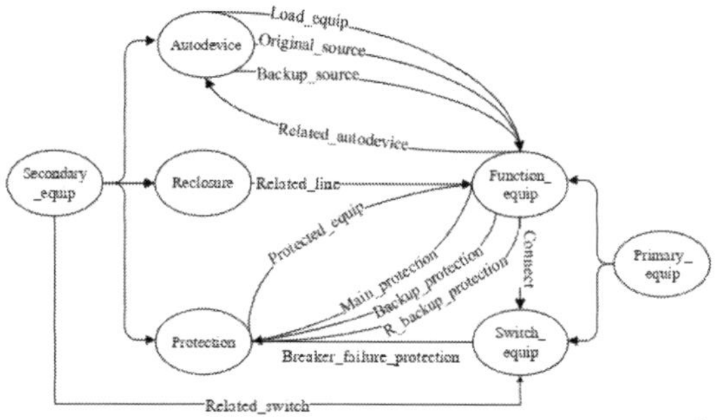

Fig. 2. Classification of electric network equipment.

In summary, the model definitions of equipment types, equipment relationships and equipment attributes in the power grid are given. Based on these, the grid state and equipment actions can be accurately described in the fault situation, which constitute the basic knowledge models of grid accident diagnosis.

3 Fault chain reasoning considering uncertainty

When the fault occurred in grid, the relative data would be uploaded to the dispatch centre through Supervisory Control and Data Acquisition (SCADA) system. These data reflect the operation of primary and secondary equipment which is the basis of the fault diagnosis. However, there are usually malfunction of relative equipment and transmission of fake data after original fault happen. The uncertainty makes the fault diagnosis difficult. Based on the fault diagnosis knowledge model above and action logic of the equipment in the fault, this paper achieves the diagnosis under uncertainty by the comparison between the fault data and the fault chain which is deduce from original fault assumption.

3.1 The basic principle of reasoning

Combined with the knowledge model in section 2, the action logic of the relative equipment is shown as figure 3. The grid state after fault is deduced considering the functional state (fault or normal) of each relative equipment. In order to improve the efficiency of diagnosis, the action logic of equipment within double faults is mainly considered and triple or more equipment faults in the actual power grid is not considered in most cases which rarely occurs and would increase the logic complexity and uncertainty. The red field in figure 3 is the equipment state assumption, and the blue field is the device action response.

And according to the logic graph, the reasoning rules are expressed in form of predicate logic which is shown in table 2.

https://doi.org/10.1051/e3sconf/202018202006

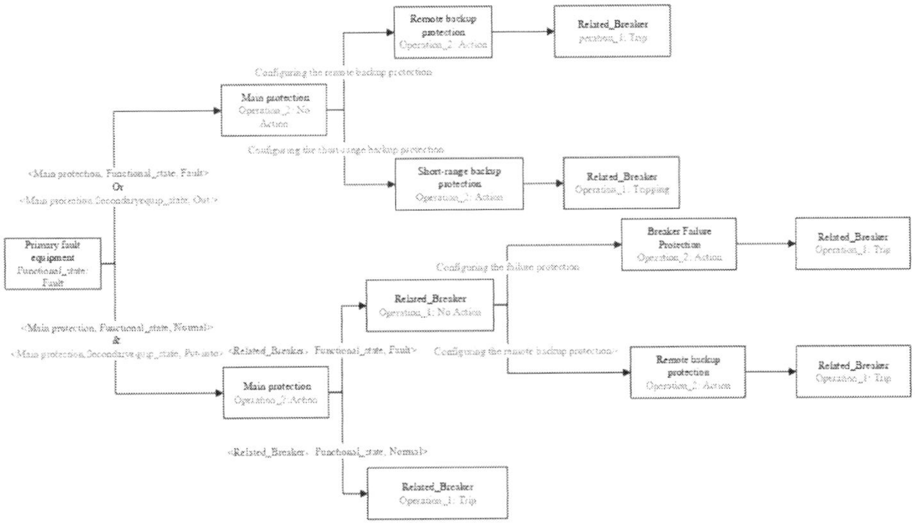

Figure 3. The action logic of relative protection and breakers after primary equipment fault

Table 2. The example of the rule expression

Sequence	Rules condition	Rules conclusion
1	∃x, y.(Primary_equip (x)∧ Functional_state (x) = Fault∧Main_protection(x, y)∧Secondary equip_state (y) = On ∧ Functional_state(y) = Normal)	Operation_2(y) = Action
2	∃x, y.(Protection (x) ∧ Operation_2(x) = Action ∧ Breaker(y)∧Related_breaker(x, y)∧ Functional_state(y) = Normal)	Operation_1(y) = Trip
3	∃x, y, z. (Primary_equip (x)∧ Functional_state (x) = Fault ∧ Main_protection (x, y) ∧ Functional_state(y) =Fault ∧ Short-range backup protection(x, z))	Operation_2(y) = No-Action Operation_2(z) = Action
4	∃x, y, z.(Protection (x)∧Operation_2(x) = Action ∧ Breaker(y) ∧ Related_breaker(x, y) ∧ Functional_state(y) = Fault ∧ Breaker Failure Protection (y, z))	Operation_1(y) = No-Action∧ Operation_2(z) = Action

After fault happened, the operation and state of the equipment in fault domain were got by analysing the data in SCADA. And then combined with the relations between the equipment which are defined in section 2, the fact base is generated. The original fault assumption set are constituted by the primary equipment in voltage loss zone. And based on the fault development logic, the fault chain reasoning is carried out step by step from original fault assumption. The reasoning results of each step are compared with the fact database to judge the validity of process of reasoning and the truth of the data. The comparison rules are as follows:

If the actions or state attributes of two devices in the fact base are logically sequential, it is confirmed that relevant information is reported correctly and the latter device has normal functional attributes; For example, if a protective action is obtained by analysing teleportation message and the corresponding switch trip occurs, and the time of both actions meets the constraint, it is confirmed that both messages are reported correctly and the switch function is normal.

1) if the state or action attribute information of a certain device is obtained by analysing the telephony information, but there is no other device action or state information logically associated with it in the fact base, the information is considered to be a false alarm and the device has not actually acted.

2) if the action or state attributes of the device obtained by accident chain reasoning are consistent with

the fact base, the relevant hypothesis is confirmed to be correct;

3) if there is no action or state information corresponding to the inference result in the fact database, but there is action or state information of upper and lower level devices logically related to the fact database, the relevant attribute information of the device is considered not reported and the relevant hypothesis is correct, otherwise, the inference is considered wrong;

4) for the complete accident chain obtained by inference, if the device action information in the fact base is not included in the accident chain and the device action information cannot be checked, it is considered that the device information is misreported; if other device information is checked with each other, it is considered that the relevant device acted wrong under the accident chain;

3.2 Choice of inference result

If multiple accident chains satisfying logical constraints are obtained through accident diagnostic reasoning, but the probability of occurrence of faults is often quite different. Here, the most probable cause of fault is selected as the diagnosis result by calculating the probability of occurrence of faults in each accident chain. According to the above analysis, the fault categories are statistically as follows. Protection fault; C. Standby self-switching fault; D. Reclosing fault; It is assumed that the occurrence probability of the above mentioned faults is P_a、P_b、P_c and P_d respectively, and the number of the corresponding fault devices in the accident chain is n_a、n_b、n_c and n_d respectively, then the calculation formula of the occurrence probability of the accident chain is as follows:

$$P_{chain} = \prod_{i=a}^{d} P_i^{n_i} \quad (1)$$

Since all values of Pi are relatively small, in order to facilitate calculation and comparison, logarithm of both sides of equation (1) can be obtained as follows:

$$\ln P_{chain} = \sum_{i=a}^{d} n_i \ln P_i \quad (2)$$

$$E_{chain} = -\ln P_{chain} \quad (3)$$

Here, the accident chain with the minimum E_{chain} value is selected as the output result of fault diagnosis.

4 Case verification

In order to prove the validity of the above diagnostic principle, a 110kV regional grid is taken as an example to illustrate the fault diagnosis process. As shown in Figure 4, breaker 123 of line JPY is equipped with distance protection, zero-sequence protection and automatic reclosing. The main transformers of Sanzhi Substation and Nanpu Substation are equipped with differential protection, gas protection and backup protection. And the 10kV bus of each substation has automatic bus transfer.

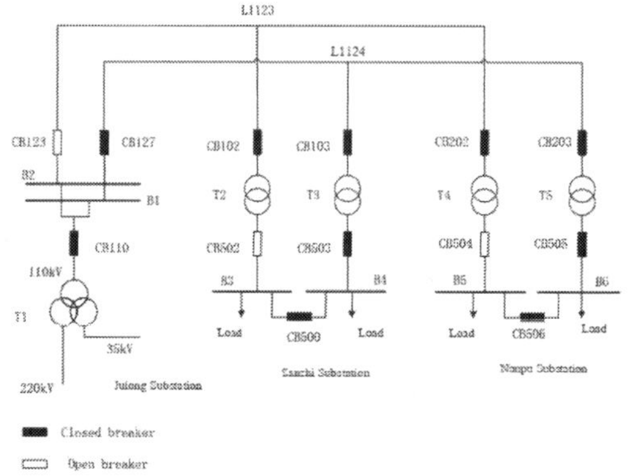

Figure 4. Structural Chart of Fault Power Network

After the fault occurs, the SCADA system receives the fault alarm messages which is shown in table 3:

Table 3. Alarm message of the fault

Substation	Equipment	Operation
Sanzhi	differential protection of T2	Action
Sanzhi	Breaker CB502	Trip
Sanzhi	automatic bus transfer equipment of B3 and B4	Action
Sanzhi	Breaker CB500	Closing
Julong	Zero-sequence II protection of L1123	Action
Julong	Breaker CB123	Trip
Julong	Reclosure of CB123	Action
Julong	Breaker CB123	Closing
Julong	Breaker CB123	Trip
Nanpu	automatic bus transfer equipment of B5 and B6	Action
Nanpu	Breaker CB504	Trip
Nanpu	Breaker CB506	Closing
Nanpu	Zero-sequence overcurrent protection of T5	Action

According to the outrage area after the fault, it is found that the fault primary equipment include line1123, transformer T2 in Sanzhi substation and transformer T4 in Nanpu substation. The initial fault hypothesis set is composed of equipment. Then the fault chain is reasoned separately. The results of the reasoning are shown as table 4:

Table 4. Results of Fault Chain Reasoning

sequence	fault chain	faulty equipment
1	Transformer T2 fault → differential protection of T2	T2

	action→ CB102 refused to move、CB502 trip → remote backup protection of T2（zero sequence II protection of L1123）action→ CB123 trip → CB123 unsuccessful reclosing，L1123 outrage → B3、B5 outrage → Automatic bus transfer equipment action → CB504 trip，CB500、CB506 close →B3、B5 recovery power supply	
2	L1123 fault → main protection of L1123 refused to act → backup protection of L1123 act（zero sequence II protection）→ CB123 trip → CB123 unsuccessful reclosing，L1123 outrage → B3、B5 outrage → Automatic bus transfer equipment action → CB504、CB502 trip，CB500、CB506 close →B3、B5 recovery power supply	L1123

As shown in the table above, there are two possible fault chains obtained by diagnostic reasoning. For the chain 1, it does not cover the protection information in Nanpu and there is no other alarm message associated with the action of the protection. So, the protection information is identified as wrong alarm message. For the accident chain 2, except for the wrong information in chain 1, the alarm message about differential protection of T2 was not covered in the accident chain either, but it was verified to be correctly reported according to the action of CB502, and CB102 was identified as refusing to act because of its closed state.

After comparing the fault equipment in the two accident chains, it can be found that the fault equipment in chain 1 is included in the fault equipment set of chain 2. According to the above formula for calculating the probability of fault, it can be inferred that the probability of occurrence of chain 1 must be higher than that of chain 2. Therefore, the final output result is that Sanzhi #2 main transformer is faulty, the main protection of the transformer action and Breaker102 refused to open. The fault is removed by the remote backup protection (zero-sequence II protection of line JPY). The information about the action of Nanpu #1 main transformer zero-sequence overcurrent protection is a fake alarm message. It can be seen that the fault diagnosis result is consistent with the manual judgment.

5 Conclusion

Through the summary of the development process of grid faults, this paper establishes a knowledge representation model for fault diagnosis. A knowledge representation method based on first-order predicate logic is provided by defining the equipment type, equipment attribute and equipment relationship which is used to describe fault characteristics and inference logic in fault diagnosis. And On the basis, summarize the logical relationship of equipment actions after the fault, as a rule, perform forward reasoning of the accident chain and check the message information of the faulty equipment, and determine the false alarm, unreported information, malfunction, and refusal equipment under the fault condition. Not only the accident diagnosis under uncertainty is realized, but also the evolution process of the fault is well reflected. The corresponding explanation of the equipment action in the fault range is given. Finally, the effectiveness of the diagnostic method is verified by the actual grid fault case.

References

1 Park Y M，Kim G W and Sohn J M 1997. A logic based expert system (LBES)for fault diagnosis of power system . IEEE Transaction on Power Systems. 12 pp 363-369.

2 Fukui C and Kawakami J 1986. An expert system for fault section estimation using information from protective relays and circuit breakers . IEEE Transactions on Power Delivery.1 pp 83-90.

3 Wenxin Guo, Fushuan Wen and Gerard Ledwich 2010. An Analytic Model for Fault Diagnosis in Power Systems Considering Malfunctions of Protective Relays and Circuit Breakers. IEEE Transactions on Power Delivery.25 pp 1393-1401.

4 Y. Zhang, C. Y. Chung, F. Wen and J. Zhong 2016 Analytic Model for Fault Diagnosis in Power Systems Utilizing Redundancy and Temporal Information of Alarm Messages, IEEE Transactions on Power Systems.31, pp 4877-4886.

5 Yongli, Z., H. Limin and L. Jinling 2006 Bayesian Networks-Based Approach for Power Systems Fault Diagnosis. IEEE Transactions on Power Delivery, 21 pp 634-639.

6 Cai, B., L. Huang and M. Xie 2017 Bayesian Networks in Fault Diagnosis. IEEE Transactions on Industrial Informatics. 13 pp 2227-2240.

7 W. xiaoli, C. guangju, X. yue and G. zhaoxin 2007 Proc. Int. Conf. on Electronic Measurement and Instruments, Fault Detection and Diagnosis Based on Time Petri Net,

8 L. Gong, R. Ma, H. Yang and Y. He 2015 Proc. Int. Conf. on Electric Utility Deregulation and Restructuring and Power Technologies, The substation fault diagnosis method based on the time constraint probability Petri net,

9 Thukaram D，Khincha P and Vijaynarasimha H P 2005 Artificial neural network and support vector machine approach for locating faults in radial distribution systems[J] . IEEE Transactions on Power Delivery 20 pp 710-721.

10 Xiaoming Han and Yan Chen 2011 Proc. Int. Conf. on Electronic and Mechanical Engineering and Information Technology (EMEIT), GrC and ANN based fault diagnosis method of distribution network.

Power System Load Forecasting Method Based on Recurrent Neural Network

Chuanjun Pang[1], *Tie* Bao[1*], and *Lei* He[1]

[1] NARI Group Co., Ltd. (State Grid Electric Power Research Institute Co., Ltd.), China, 100192

Abstract. Power system load forecasting plays an important role in the power dispatching operation. The development of the electricity market and the increasing integration of distributed generators have increased the complexity of power consumption model and put forward higher requirements for the accuracy and stability of load forecasting. A load forecasting method based on long-short term memory (LSTM) is proposed. This method uses deep recurrent neural network from the artificial intelligence field to establish a load forecasting model. Using the LSTM network to memorize the long-term dependence of the sequence data, the intrinsic variation of the load itself is identified from both the horizontal and vertical dimensions within a longer historical time period, while considering various influencing factors. Actual load data is used to verify the forecasting performance of different historical date windows and different network architectures.

1 Introduction

Power system load forecasting can estimate the power demand for a period in the future. The future unit commitment, generation plan, tie line exchange plan and organization of the electric power transactions can be arranged according to the load forecast result. Therefore, load forecasting is the basis to realize the safe, reliable and economical operation of power system, and it is also an important guarantee to improve the utilization ratio of power generation equipment and the efficiency of economic dispatch. Accurate load forecasting has great significance to optimal unit commitment, economic dispatch, optimal power flow and electricity market transaction, etc [1].

Power system load is affected by economic factors, weather, season, date, and other factors, so load variation has random components. With the development of power system, the types of user load and influencing factors are increasing. In the electric power market environment, the complexity of electricity consumption mode is increased by the users' behaviour, the response of users to the incentive policies and the urban development situation. The introduction of large-scale distributed generators and the widespread use of electric vehicles have increased the load fluctuation [2]. Peak-valley time-of-use price through its own economic leverage, urges users to constantly change the way they use electricity [3], and stimulates and encourages electric vehicles to change their charge-discharge behaviour [4].

At present, load forecasting methods are mainly divided into two categories: statistical methods and artificial neural network (ANN) based methods. Statistical methods use time series prediction to forecast [5], mainly including multiple linear regression (MLR), auto regression (AR) and autoregressive moving average (ARMA). The models of such methods are simple, but only a small number of influencing factors and sample data can be processed, and the stability of the original time series is highly demanded, because these methods are based on the assumption that the historical load, load influencing factors and future load are linearly related. In fact, the influencing factors and the load changes have strong non-linear characteristics. Artificial intelligence based methods include artificial neural network (ANN), support vector machine, fuzzy logic and other modern forecasting methods [6,7,8]. Artificial neural network has been successfully applied in speech recognition, natural language processing, computer vision and other fields [9,10,11]. Many researchers have tried to apply this method to load forecasting and new energy power forecasting and achieved good results [12,13,14]. ANN simulates brain mechanisms, learns implicit rules from data and forecasts the future data. ANN has a strong nonlinear modelling ability and can learn the nonlinear characteristics of the load from the historical load data. ANN includes a variety of different network architectures. At present, feedforward neural network (FNN) is widely used in power system load forecasting [14]. But FNN can't consider the dependence between the training samples, can't model the sequence data, and can't establish the load forecasting model from a longer historical time scale.

In this paper, an improved recurrent neural network (RNN) of long-short term memory unit is used to establish a load forecasting model. Using the deep recurrent neural network's characteristics of remembering historical information, based on many historical load data,

* Corresponding author: bt0504@126.com

the inherent rules of load variation can be identified from a longer historical time range, and the complex nonlinear relationship between the load influencing factors and the load can be learned. Actual load data are used to verify the model. The experimental results show that the method can identify the load variation rules to a certain extent and make a more accurate forecast of the future load.

2 Recurrent neural network

2.1 Ordinary recurrent neural network

Recurrent neural network (RNN) is a kind of neural network which can process sequence data [15]. The network architecture is shown in Figure 1.

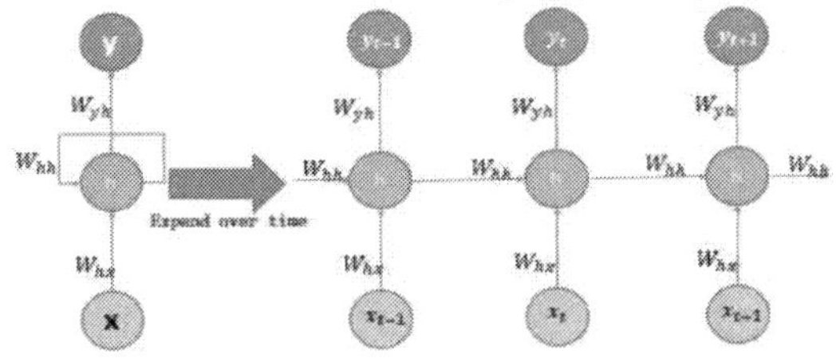

Figure 1. Recurrent neural network.

Compared with FNN, each neuron in RNN hidden layer can be recycled through self-link. Given the input data $X = \{x_1, x_2, x_3, \ldots x_T, \}$, RNN uses formula (1) and formula (2) to calculate hidden state $h = \{h_1, h_2, h_3, \ldots h_T, \}$ and output $y = \{y_1, y_2, y_3, \ldots y_T, \}$.

$$h_t = f(W_{hx}x_t + W_{hh}h_{t-1} + b_h) \quad (1)$$

$$y_t = g(W_{yh}h_t + b_y) \quad (2)$$

W_{hx}, W_{hh}, W_{yh} represent input weight, hidden weight and output weight, respectively. f, g represent hidden layer activation function and output layer activation function, respectively. RNN uses hidden state to remember all previous input information. The output y_t at the current time is not only affected by the current input x_t, but also by the hidden state h_t at the previous time. RNN uses back-propagation through time to train model [12]. In the training process, RNN has difficulty in learning long-distance dependence because of the disappearance of gradient.

2.2 Long-short term memory

Long-Short Term Memory (LSTM) is an improvement on ordinary RNN. The long-term dependence problem, which makes the model not suitable on learning sequence data because of gradient disappearance, can be solved through introducing memory cell into each neuron in hidden layer, and using forget gate, input gate and output gate to control the state of memory cell [15]. The architecture of LSTM neuron is shown in Figure 2:

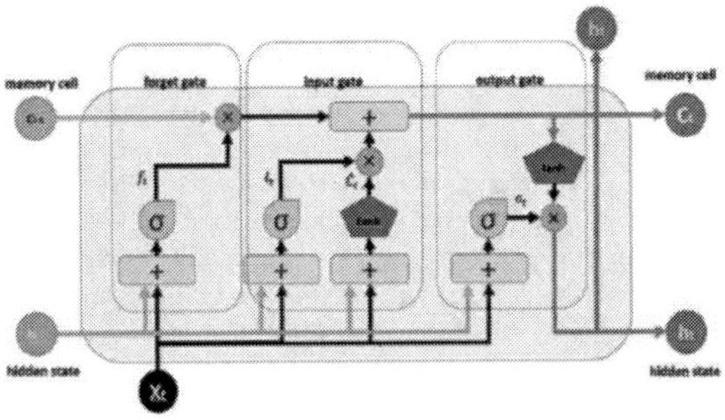

Figure 2. The architecture of LSTM.

Memory cell and the hidden state can memorize the historical information of the sequence data together. The information in the memory cell is controlled by three gate units.

Forget gate deletes the information in the memory cell according to the hidden state of the previous moment h_{t-1} and the input of the current moment x_t. Forget gate is calculated as follows:

$$f_t = \sigma(W_f[h_{t-1}, x_t]) + b_f \quad (3)$$

Input gate adds information to the memory cell according to the hidden state of the previous moment h_{t-1} and the input of the current moment x_t. The functions are shown in formula (4) and formula (5). i_t represents the information that needs to be memorized; $\hat{C}_t$ represents the candidate memory cell to update next memory cell.

$$i_t = \sigma(W_i[h_{t-1}, x_t]) + b_i \quad (4)$$

$$\hat{C}_t = \tanh(W_c[h_{t-1}, x_t]) + b_c \quad (5)$$

After calculating the forget gate and the output gate, formula (6) is used to update the memory cell.

$$C_t = f_t * C_{t-1} + i_t * \hat{C}_t \quad (6)$$

Output gate decides the current hidden state h_t according to the hidden state of the previous moment h_{t-1}, the input of the current moment x_t and updated memory cell C_t, as shown in formula (7) and formula (8).

$$o_t = \sigma(W_o[h_{t-1}, x_t]) + b_o \quad (7)$$

$$h_t = o_t * \tanh(C_t) \quad (8)$$

In the formulas, W_f represents forget gate weight; W_i, W_c represent input gate weights; W_o represents output weight; b_f, b_i, b_c, b_o represent corresponding offset vectors.

3 Load forecasting based on LSTM

3.1 Basic theory

Power system load is periodic because of the influence of people's production and living patterns, and it also has great uncertainty because of the influence of random factors such as weather changes, major events and so on. The load l_t^d at a certain time of a day is affected by the type of the day (days of the week, whether it is a holiday or not) and the current weather factors (temperature, humidity, etc.). It is also related to the load l_{t-n}^{d} at the past time of the day and the load l_t^{d-w} at the same time of the past few days, where t is the moment, d is the date, and w is the historical date window. LSTM has the advantage of learning the rules of sequence data from a long-time range. It can recognize the load variation regularity of the forecasting day horizontally, identify the load variation regularity at the same time in the historical date window vertically, and consider the influence of related factors on the load of the forecasting day.

3.2 Forecasting model

Based on LSTM network, the forecast day load curve is forecasted using the load curves of w days before forecast day and the influencing factors of forecast day. The LSTM network structure is shown in Figure 3.

$$\hat{L}^d = \{\hat{l}_0^d, \hat{l}_1^d \dots \hat{l}_T^d\} \quad (9)$$

$\hat{L}^d$ is the forecasted load for forest day, and T is called step size, T=24|96|288|1440.

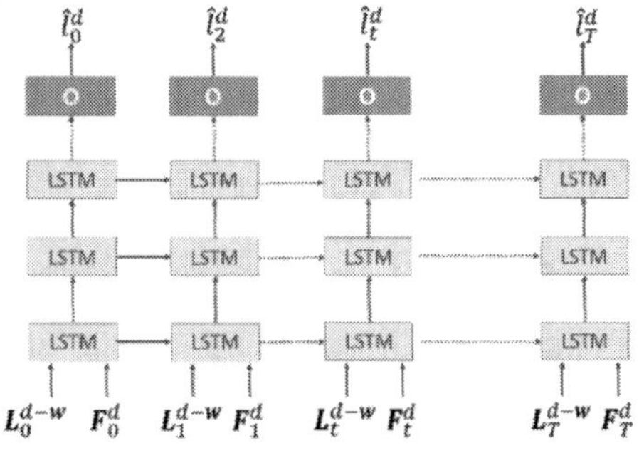

Figure 3. LSTM neural network model.

$$L_t^{d-w} = \{l_t^{d-1}, l_t^{d-2} \dots l_t^{d-w}\} \quad (10)$$

l_t^{d-w} represents the load at time T of w days before forecast day. w is the historical date window.

$$F_t^d = \{month_t^d, is\_holiday_t^d, \dots, temp_t^d\} \quad (11)$$

It includes load influencing factors, such as date types, holidays, weather and so on.

The input of the load forecasting model at time T is $input = \{L_t^{d-w},\ F_t^d\}$ and the output is $output = \hat{l}_t^d$.

3.3 Model training

Model training includes three stages: data pre-processing, model training and model evaluation.

3.3.1 Data preprocessing

Data preprocessing mainly includes two steps: Vectorization and normalization. Neural network is based on linear algebra theory, and cannot be trained directly on the original data. So it is necessary to convert the original data into vector before training. Data vectorization concatenates the load data in the history date window L_t^{d-w} with the influence factor data of the forecast day F_t^d, and transforms the concatenated data into vector.

The neural network is trained by the back-propagation algorithm based on gradient descent. If the data is too large or too small, it will be difficult to find the optimal solution in the training process. Therefore, normalization of data into a standard interval is conducive to the solution of the model. The min-max scaling normalization is used to normalize each element of the vector into the interval [0-1], as shown in formula 12 [16]:

$$X_{norm} = \frac{X - X_{min}}{X_{max} - X_{min}} \quad (12)$$

where X_{max} is the maximum value, X_{min} is the minimum value, X is the original value and X_{norm} is the normalized value.

3.3.2. Model training.

Back-propagation through time (BPTT) [17,18] algorithm is used to train the LSTM network for load forecasting. The training objective is to adjust the network parameters so that the network output is as close to the real values as possible. The training process is as follows:

1) Initialize network parameters.

2) Input the training data to the network and calculate the network output.

3) Calculate the difference between the network forecast value and the actual value, and use loss function to measure the error.

4) According to the error, calculate the gradients of network parameters over loss functions in two directions of network level and time.

5) Adjust network parameters according to the gradient.

6) Repeat steps 2-5.

Steps 2-3 constitute the forward reasoning process, and steps 4-5 constitute the backward propagation process.

The loss functions include mean square error (MSE), mean absolute error (MAE), mean absolute percentage error (MAPE) and so on. In this paper, MSE is used as the loss function, which is defined as follows [18]:

$$Loss(W, b) = \frac{1}{N}\sum_{i=1}^{N}(\hat{L}_i - L_i)^2 \quad (13)$$

where N represents total number of training samples, $\hat{L}_i$ represents forecasted load, and L_i represents actual load.

3.3.3 Model evaluation.

In the process of model training, all data are used to input into the network to adjust its parameters, which is called an epoch. If the number of epoch is too small the network will be under-fitting; if the number of epoch is too large the network will be over-fitting. In order to determine the appropriate epoch and select the optimal model, the model should be evaluated during each epoch. The following steps are used to determine the optimal epoch.

1) Divide historical load data into training set, validation set and test set.

2) Input the training set into the network for training. After an epoch, the validation set is put into the network model, and the model is evaluated by formula (13).

3) If the performance of the current model exceeds those of all the previous epochs, keep the model as the optimal model.

4) Repeat steps 2-3 until the exit condition is satisfied. (The performance of continuous N epochs can't be improved any more.)

5) Output the optimal model and use test set to test the model.

4 Example analysis

4.1. Experimental data

The hourly historical system load data of Electric Reliability Council of Texas (ERCOT) control area from 2003 to 2016 are used to validate the proposed model [18]. Data from 2003 to 2014 are used as training set, data from 2015 as validation set and data from 2016 as test set. Validation set is used to verify the optimal number of training epoch, the optimal network structure and parameters. Test set is used to test the forecast accuracy of the model. While the historical weather information is not available, the weather factors are not taken into account in the model training process. Only the type of date (working day, the days of week), the same time load in the historical window period and maximum and minimum daily load in the historical window period are used.

4.1.1 Evaluation indices

The common model evaluation indices include mean square error (MSE), mean absolute error (MAE), mean absolute percentage error (MAPE). The definition of MSE is shown in formula (13). MAE [18] and MAPE [1] are defined as in formula (14) and (15).

$$MAE = \frac{1}{N}\sum_{i=1}^{N}\left|\hat{L}_i - L_i\right| \quad (14)$$

$$\mathrm{MAPE} = \frac{1}{N}\sum_{i=1}^{N}\left|\frac{\hat{L}_i - L_i}{L_i}\right| \quad (15)$$

N represents total number of training samples, $\hat{L}_i$ represents forecasted load, and L_i represents actual load.

4.2. Network structure

In LSTM network, the number of hidden layers and the number of LSTM neurons in each hidden layer affect the accuracy of load forecasting. Using the hourly historical load data of 2015 as validation set, with fixed historical date window $w = 7$, the number of hidden layer neurons is selected layer by layer through enumeration method [1] to determine the optimal network structure. First, the optimal number of hidden neurons in the first layer is determined and fixed; then a hidden layer is added to determine the optimal number of hidden neurons in the second layer. The process is continued until the forecast accuracy is no longer improved.

The forecasting performance under different network structures is shown in Table 1. When the number of hidden neurons in each layer is selected, options are 5-40 with interval 5 and 8 levels. The number of hidden layers is set to 1, 2 and 3 in turns. As shown in Table 1, when the hidden layer is 1, the number of hidden neurons in the layer is 5, the minimum MAPE is 4.6%; when the hidden layer is 2, the number of hidden neurons in each layer is 20, the minimum MAPE is 4.72%; when the hidden layer is 3, the number of hidden neurons in each layer is 10, the minimum MAPE is 4.58%. Therefore, when $w = 7$, the number of hidden layer is 2 and the number of hidden neutrons in each layer is 10, the forecasting performance of the network is better.

Table 1. Forecasting performance of LSTM model with different structures

Hidden layer	Hidden cells	MAE/MW	MAPE/%	Optimal epoch
1	5	1772.27	4.6	500
	10	1800.32	4.72	38
	15	1792.25	4.74	53
	20	1844.75	4.94	40
	25	1815.24	4.79	30
	30	1878.15	4.99	24
	35	1808.85	4.80	35
	40	1852.92	4.80	16
2	5	1813.37	4.75	172
	10	1905.45	4.99	57
	15	1822.30	4.77	38
	20	1801.05	4.72	42
	25	1798.51	4.71	35
	30	1991.26	5.11	20
	35	1800.72	4.83	16
	40	1831.83	4.75	16
3	5	1782.06	4.59	180
	10	1779.85	4.58	65
	15	1824.71	4.74	48
	20	1785.96	4.65	55
	25	2002.88	5.14	20
	30	1790.58	4.65	52
	35	1915.15	5.00	20
	40	1891.02	4.87	39

4.3. Historical date window

Set the network layer as 1, 2, 3 and the number of hidden cells as 5, 20, 10, respectively. The historical window w takes 1-60 days. The hourly historical load data of 2015 are used as validation set.

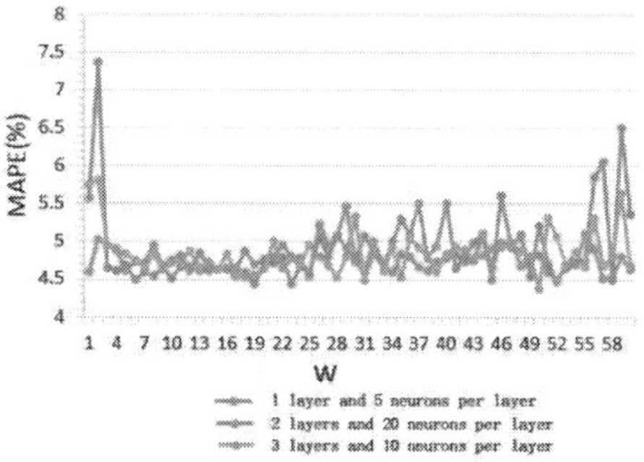

Figure 4. The change of MAPE with **W** under different LSTM network structures.

Figure 4 shows the curves of MAPE changing with ***w*** under different network structures. As shown in Figure 4, with the increase of ***w***, MAPE of different network structures are almost the same. There is no obvious downward or upward trend, but the change of MAPE becomes unstable when ***w*** increases. Especially when the network has 1 layer and there are 5 neurons in the layer, the larger ***w*** causes MAPE to change violently. Under three network structures, when the value of ***w*** is in [7,30], MAPE is relatively low and stable; when the hidden layer is 1, the number of hidden neurons in the layer is 5 and ***w*** is 23, the forecasting performance is the best and MAPE is 4.4%.

4.4. Result analysis

In order to verify the validity and scientificity of the proposed method, the optimal forecasting model (hidden lay is 1, the number of neurons is 5 and ***w*** is 23) and the test set (hourly load of year 2016) are used to test the forecasting performance of the model.

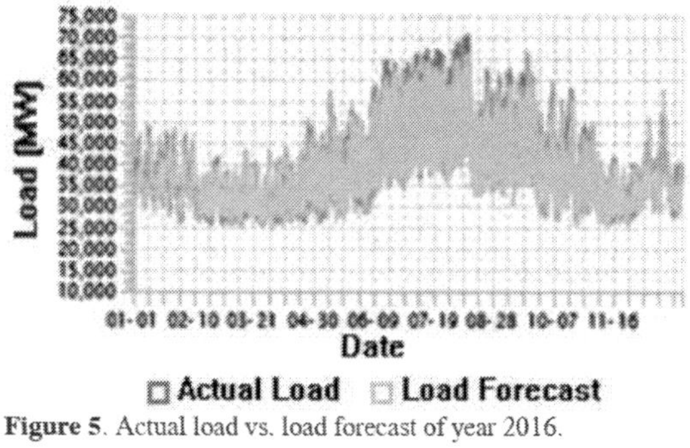

Figure 5. Actual load vs. load forecast of year 2016.

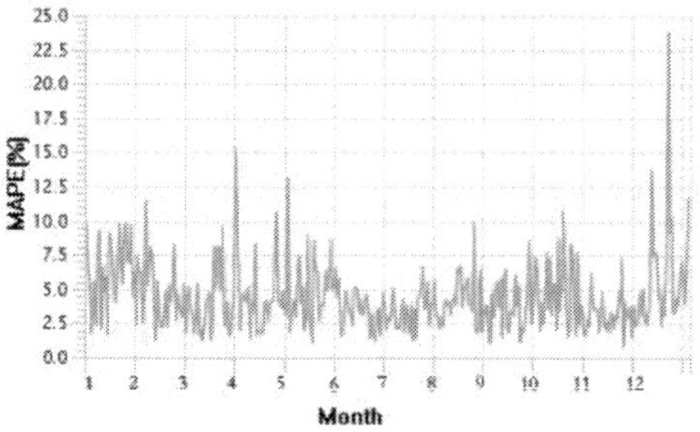

Figure 6. Daily forecasting MAPE.

The forecast results are shown in Figure 5, and the daily MAPE is shown in Figure 6, with average MAPE 4.45%.

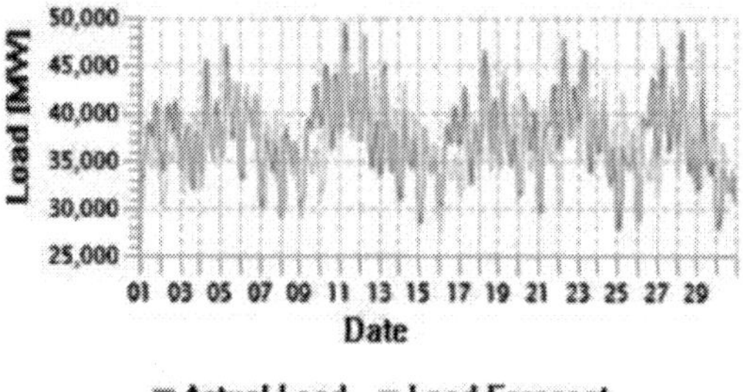

Figure 7. Actual load vs. load forecast of Jan, 2016.

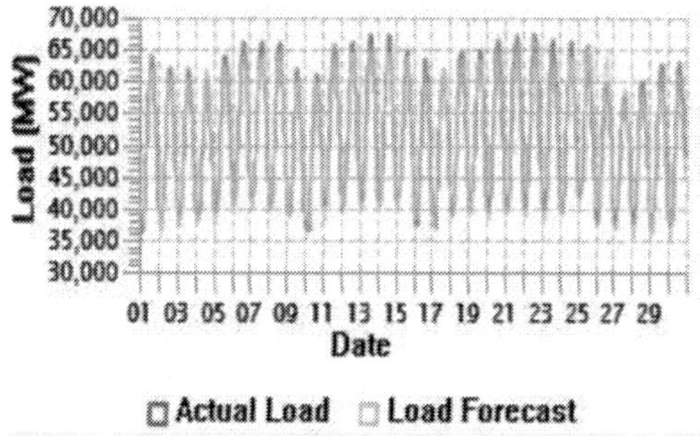

Figure 8. Actual load vs. load forecast of July, 2016.

Figure 7 and Figure 8 show the comparison between actual load and load forecast in 2016, January and July, respectively. The analysis shows that system load changes in January have great randomness, because the model does not consider weather, holidays (except Saturday and Sunday), special events and other factors. So the forecasting accuracy is low and only based on load itself, and MAPE is 6.66%. The forecasting accuracy of July is higher except the daily peak load and valley load, and MAPE is 3.51%.

According to the forecasting results, the forecasting model used in this paper can learn the variation of the historical load and accurately forecast the future load.

5 Conclusion

In this paper, recurrent neural network (RNN), from the field of artificial intelligence, is improved with long-short term memory (LSTM) unit to build load forecasting model, taking the advantage that LSTM can learn the long-term dependence of sequence data to identify the variation rules of load itself. The actual load data are used to verify the model and the influence on forecasting performance from different neural network structures and different historical time windows. The experimental results show that the LSMT network can forecast the future load accurately.

In the LSTM network training, the influence of historical weather factors, major events and other random factors on the load is not considered. The next step is to consider more affecting factors to validate the model and further improve the forecasting accuracy of the model.

Artificial intelligence technology, represented by deep neural networks, has made remarkable achievements in many fields. In the future, recurrent neural network (RNN) will be applied to more business areas related to power system dispatching operation, such as wind and photovoltaic power forecasting, voice dispatching log, dispatching robot and so on, to improve the intelligent level of power system dispatching operation.

This work is supported by the Science and Technology Project <Research on the technology of new energy brought into the medium- and long-term power balance considering the fluctuation characteristics of output time series (4000-201940193A-O-O-OO)> from the State Grid Corporation of China.

References

1. KONG Xiangyu, ZHENG Feng, E Zhijun, et al, "Short-term Load Forecasting Based on Deep Belief Network[J]", Automation of Electric Power Systems, 2018, 42(5), pp. 133-139.
2. LIUJ, GAO H,ZHAO M A, et al, "Review and prospect of active distribution system planning[J]", Journal of Modern Power Systems and Clean Energy, 2015, 3(4), pp. 457-467.
3. YAO Jiangang, FU Qiang, YE Lun, et al, "Substation Capacity Planning Method Considering Influence of Peak-valley Time-of-use Power Price[J]", Automation of Electric Power Systems, 2017, 41(13), pp. 53-61.
4. LIU H, ZENG P,GUO J, et al, "An optimization strategy of controlled electric vehicle charging considering demand side response and regional wind and photovoltaic[J]", Journal of Modern Power Systems and Clean Energy, 2015, 3(2), pp.232-239.
5. LI Bo, MEN Deyue, YAN Yaqin, et al, "Architecture and Key Technologies for Auxiliary Decision-making System of Integrated Dispatch Scheduling[J]", Automation of Electric Power Systems, 2015, 39(1), pp.137-140.
6. WU Qianhong, GAO Jun, HOU Guangsong, et al. "Short-term Load Forecasting Support Vector Machine Algorithm Based on Multi-source Heterogeneous Fusion of Load Factors[J]", Automation of Electric Power Systems, 2016, 40(15), pp. 67-72.
7. WU Xiaoyu, HE Jinghan, ZHANG Pei, et al, "Power System Short-term Load Forecasting Based on Improved Random Forest with Grey Relation Projection[J]", Automation of Electric Power Systems, 2015, 39(12), pp. 50-55.
8. XIAO Bai, NIE Peng, MU Gang, et al, "A Spatial Load Forecasting Method Based on Multilevel Clustering Analysis and Support Vector Machine[J]", Automation of Electric Power Systems, 2015, 39(12), pp. 56-61.
9. Lipton, Zachary C., Berkowitz, John, and Elkan, Charles, "A critical review of recurrent neural networks for sequence learning", arXiv preprint arXiv:1506.00019, 2015.
10. Justin Bayer, Daan Wierstra, Julian Togelius, and J'urgen Schmidhuber, "Evolv-ing memory cell structures for sequence learning", Artificial Neural Networks–ICANN 2009, pp. 755–764, Springer, 2009.
11. Yoshua Bengio, Patrice Simard, and Paolo Frasconi' " Learning long-term depen-dencies with gradient descent is difficult", Neural Networks,IEEE Transactions on, 5(2), pp. 157–166, 1994.
12. XU Longbo,WANG Wei,ZHANG Tao, et al, "Ultra-short-term Wind Power Prediction Based on Neural Network and Mean Impact Value[J]", Automation of Electric Power Systems, 2017, 41(21), pp. 40-45.
13. LIU Ruiye, HUANG Lei, "Wind Power Forecasting Based on Dynamic Neural Networks[J]", Automation of Electric Power Systems, 2012, 36(11), pp. 19-22.
14. HERNANDEZ L, BALADRON C, AGUIAR J M, et al, "Artificial neural network for short-term load forecasting in distribution systems[J]", Energies, 2014 ,7(3), pp. 1576-1598.
15. Bengio Y, Boulanger-Lewandowski N, Pascanu R, "Advances in Optimizing Recurrent Networks", IEEE International Conference on Acoustics, 2012, pp. 8624–8.
16. Adam Gibson, Josh Patterson, "Deep Learning : A Practitioner's Approach [M]", Boston :O'Reilly Media, 2017.
17. Wim De Mulder, Steven Bethard, and Marie-Francine Moens, "A survey on the application of recurrent neural networks to statistical language modeling", Computer Speech & Language, 30(1), pp. 61–98, 2015.
18. "Hourly Load Data Archives [EB/OL]", [2018-06-12]. http://www.ercot.com/gridinfo/load/load_hist/.

 https://doi.org/10.1051/e3sconf/202018202008

Development of DC Energy Meter for Renewable Energy Application

P N Gajera[1,*], *K S* Parikh[1] , *V* Gupta[1] and *H* Gupta[2]

[1]Electrical research and development association, Vadodara, Gujarat, India
[2]SVIT, Vasad, Vadodara, Gujarat, India

Abstract. The world we live in, is becoming more and more dependent on electrical energy, which, sadly, still heavily relies on outdated techniques involving the burning of fossil fuels. Fortunately, more and more people are accepting the fact that this system simply isn't sustainable in the longer-run. Even if air pollution wasn't a huge factor, the shortage of these resources is something that is bound to happen in the nearest future, and it would pose an impassable obstacle towards the future growth of mankind. Hence, importance of renewable energy is such an important trend for the future of mankind. At present, net AC meters are available for measurement of four quadrant energy, which is used at output of both on grid and off grid solar inverter. This meter can measure both, energy fed into the grid and energy utilize from the grid. However Energy generated by solar panel cannot be measured directly. Even in off grid inverter the present net meter cannot measure energy generated by solar panel. In case of DC micro grid, where DC energy is used/ consumed directly by the load, the present net meter cannot be used as they can be used in AC energy measurement only. With the launch and implementation of national energy saving and emission reduction policies in relation to distributed power supply by renewable power generation, the DC energy metering technology will become an inevitable trend, breaking the landscape dominated by AC in the field of electric energy metering. This paper describes introduction, block diagram [1], Hardware and software description for development of dc energy meter.

1 Introduction

In today's world, two of the biggest challenges are climate change and management of diminishing fossil fuel reserves. In order to secure the future for ourselves and generations to come, we must act now to reduce energy consumption and substantially cut greenhouse gases, such as carbon dioxide. The consumption of energy is increasing at a fast pace while available resources remain limited. Out of the total electricity produced, over certain amount comes from fossil fuels. Coal and other fossil fuels, have taken hundreds of millions of years to form, are likely to deplete soon. In last two hundred years, we have consumed 60% of all resources. For sustainable development, we need to adopt energy efficiency measures. We need to use renewable sources for production of electricity. One such method is by using solar panels which uses solar energy for production of electricity.

To fulfil the mission of power to all, government of India is promoting DC micro grid so that local energy requirement can be fulfilled. Hence, the demand for DC metering will be increasing in renewable energy applications like solar energy panels metering, wind turbines and many more.

Moreover, now a day with the increase in renewable energy applications, DC circuits are rising in prominence. With the proliferation of DC circuits, the need for capable DC metering [2] technology has come to the fore. The total energy needed in a given time interval is known as consumption, and can be measured & data can be stored using energy meters. Energy meters measure the energy flowing through some point of an electric circuit, Kilowatt hour (kWh) is the most generally used unit for measuring and billing electrical energy, commonly known as the billing unit. Apart from DC energy, the DC energy meter will measure DC voltage, current and power and additional parameters like temp and humidity & pressure of surrounding atmosphere for renewable energy application. To better understand energy consumption, consider a 100 W lamp that remain on for 10 hours. To produce light this lamp will consume 100 W in an hour, i.e., 100 Wh. If this lamp is turned on for 10 hours, the total consumption will be

$$100 \text{ W x } 10 \text{ h} = 1000 \text{ Wh} = 1 \text{ kWh}.$$

At this point, it is clear to understand that after these 10 hours the energy meter will consider one more kWh, for which applicable electricity rate will be charged.

2 Blockdiagrm

* Corresponding author: puran.gajera@erda.org

Fig.1 shows the block diagram of the overall system of DC energy meter. Arduino Uno controller takes the data from all sensors i.e. Pressure sensor, Temperature and Humidity sensor, RTC for date and time. After the sensing of voltage and current, Ardunio controller will feed all data to the LCD 20X4.

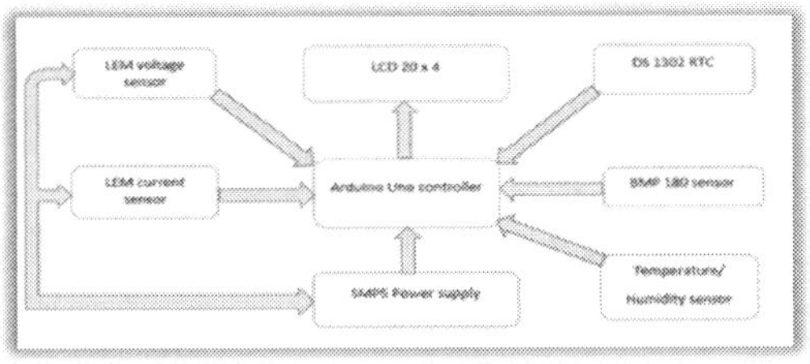

Fig. 1. Block Diagram of the DC energy meter

3 Developed system

The proposed system consists of Arduino Uno as a heart of the system, voltage sensor, current sensor, Temperature sensor, Humidity sensor, RTC for Date and Time, Pressure sensor, 20 x 4 LCD and Arduino IDE for programming.

3.1 Hardware description

3.1.1 Voltage sensor

There are various method for sensing the voltage like voltage divider network and potential transformer. Voltage divider network can be made by resistors divider or capacitors divider.

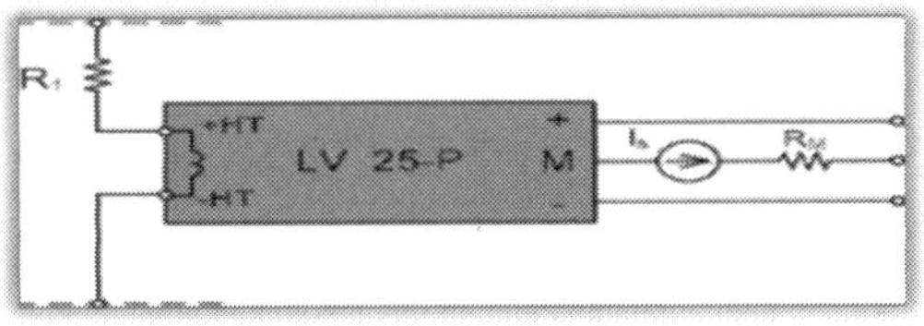

Fig. 2. Voltage Sensor

We have used LEM 25-P voltage sensor due to good linearity response & excellent accuracy features for dc voltage sensing. This sensor works on the principle of Hall Effect[3] based closed loop current transducer (Figure 2). 25 kΩ resister is connected in series with the primary circuit of the voltage transducer. 100 Ω resister is connected across secondary output and GND terminal for getting the output voltage. +12 volt & -12 volt is given by Power supply circuit to operate this voltage transducer. This sensor output is connected to Analog 0 channel of Ardunio controller. Sensor response was measured by using Fluke 5100 calibrator, different voltages were given as input of transducer and measured the respective output. After getting calibration data, linear equation was fitted in the program for displaying measured value.

3.1.2 Current sensor

LEM LA 55-P sensor (Figure 3) i.e. Hall Effect based current sensor was used for development of dc energy meter due to good linearity result & excellent accuracy features. It requires ± 12 volt for operation and need 150 Ω resistance at secondary terminal. Output of current sensor was connected with analogue channel 1 of Arduino controller. Sensor response was measured by using Fluke 5100 calibrator, different current/s were passed through input of transducer and the respective output was measured. After getting calibration data, linear equation was fitted in the program for displaying measured value.

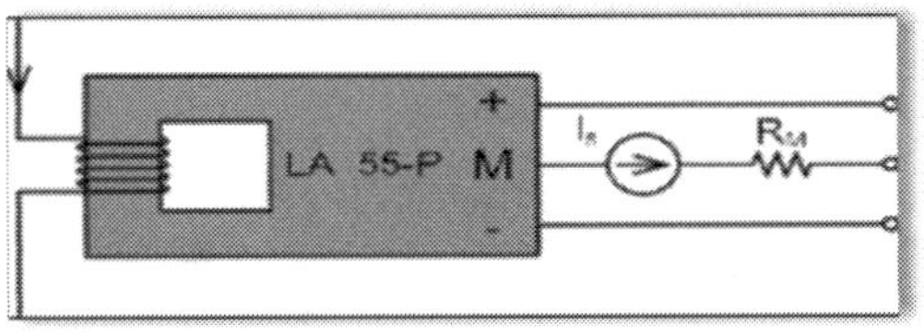

Fig. 3. Current Sensor

3.1.3 BMP180 sensor

The BMP180 (Figure 4) is a new generation of high precision digital pressure sensor for consumer applications. The ultra-low power, low voltage electronics of the BMP180 is optimized for use in mobile phones, PDAs, GPS navigation devices and outdoor equipment. With a low altitude noise of merely 0.25m at fast conversion time, the BMP180 offers superior performance. The I2C interface allows for easy system integration with a Arduino controller SCL and SDA pin of BMP180 sensor is connected with A4 and A5 pin of Arduino controller.

Fig. 4. BMP 180 sensor

3.1.4 Temperature/Humidity sensor

This sensor (Figure 5) measures Relative Humidity and Temperature and outputs in simple serial interface with 9600 baud rate. TX pin of this sensor interfaced with Arduino microcontroller's RX pin. +5V and GND pins are connected with Arduino controller's +5V and GND pins. This sensor gives the 13 character output which stored in array and sends this data to LCD as well as serial port.

Fig. 5. Temperature/Humidity sensor

3.1.5 DS1302 RTC for data and temperature

We have used DS 1302 based RTC module (Figure 6) for updating recorded all data with date and time. SCLK pin, DAT pin, RST pin of RTC modules are interfaced with 2, 3 and 4 pin of Arduino controller.

Fig. 6. DS 1302 RTC module

3.1.6 Ardunio Uno controller

Arduino Uno (Figure 7) is based on AVR microcontroller called Atmega328. This controller comes with 2KB SRAM, 32KB of flash memory, 1KB of EEPROM. Arduino Board comes with 14 digital pins and 6 analog pins. ON-chip ADC is used to sample these pins. A 16 MHz frequency crystal oscillator is equipped on the board. Arduino Uno is a very valuable addition in the electronics that consists of USB interface, 14 digital I/O pins, 6 analog pins, and Atmega328 microcontroller. It also supports serial communication using Tx and Rx pins. It is an open-source platform. The software used for Arduino devices is called IDE (Integrated Development Environment) which is free to use and required some basic skills to learn. It can be programmed using C and C++ language. Atmega328 microcontroller is placed on the board that comes with a number of features like timers, counters, interrupts, PWM, CPU, I/O pins and based on a 16MHz clock that helps in producing more frequency and number of instructions per cycle. Programming is done in such a way so that Ardunio controller senses the voltage, current and sensor data and display on LCD 20x4.

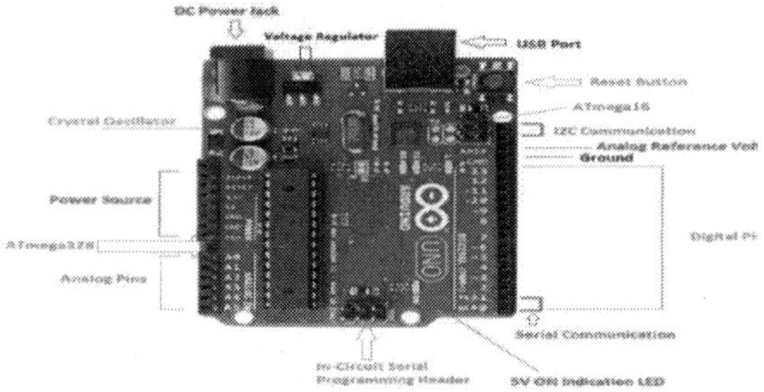

Fig. 7. Ardunio Uno controller

3.2 Software description

3.2.1 Software algorithm

Figure 8 shows the software algorithm which was used during development of DC energy meter. In this algorithm, initially we setup the initiation of RTC, LCD, UART and I2C with their pin. If I2C protocol will not give data of pressure, then "Error" message shows in serial port. During the continuous loop i.e. while loop, Arduino controller initially checks for serial data. If data is not coming then it will wait and nothing display on LCD. Once data receive from serial port, it will take the 13 bit character data of temperature/humidity sensor. It will split the particular temperature data & humidity data from 13 bit character data and send the data on particular column and row position of 20x4 LCD. After that we are calling data of BMP085 sensor i.e. pressure data which we are putting in variable and display on LCD & then we are calling DS 1302 routine for taking data of date and time. Controller receives date and time data through SPI communication and fed to LCD for display. Now controller calling time calculation routine for calculating time to complete whole while routine. Output of time calculation routine gives the required time in variable time. After completion of time routine, controller senses the voltage and current data through output of LEM voltage and current transducer through ADC of Arduino controller. DC power will be calculated through software by product of dc voltage and dc current. For calculating DC energy will be calculated by dc power and time for 1 hour. These dc power and dc energy data will be fed to the LCD using controller. After completed this routine, it will again repeat the process and update the all data on LCD.

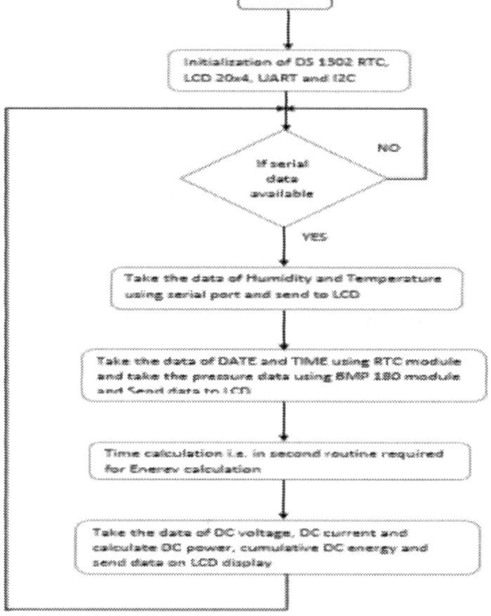

Fig. 8. Flow chart of software algorithm

3.2.2 Arduino IDE complier

The open-source Arduino Software (IDE) makes it easy to write code and upload it to the board. It runs on Windows, Mac OS X, and Linux. The environment is written in Java and based on Processing and other open-source software. Adafruit BMP085 library for interfacing BMP180 sensor to measure pressure data, DS1302 library for interfacing of RTC module to measure date and time data, EEPROM library for storing

 https://doi.org/10.1051/e3sconf/202018202008

DC energy value when power goes off and display library for data display on LCD are added in the Ardunio IDE compiler for programming.

3.2.3 GUI software

The open-source Arduino Visual basic based data logger was developed for data storing of temperature, humidity and energy data with date and time. These data are coming from Ardunio controller via serial communication before that user will select and detect comport using combo box option given in GUI software[4]. Once communication done, controller starts the sending required data for GUI and update the value. Figure 9. Shows the visual basic based developed GUI software.

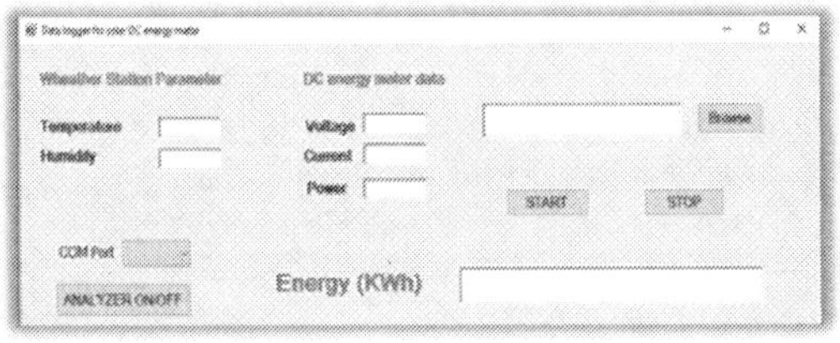

Fig. 9. GUI software

3.2.4 Experimental results

Experiment was carried out with the help of programmable dc source. Output of Programmable source was connected with input of developed dc energy meter. Variable DC voltage and dc current was set by programmable dc source and verified voltage and current value with developed dc voltage and dc current value. In this developed dc energy meter, we are continuously taking average of dc energy for 1 minutes and this updated dc energy value display on LCD. We are using ardunio controller which have 10 bit adc with LEM voltage and LEM current sensor. So we can sense low voltage value and get accuracy better. Developed energy meter have accuracy with ± 1%. Table 1 shows the achieved experiment result at 5A load and Table 2 shows the achieved experiment result at 10A load

Table 1. Energy Accuracy at 5A

DC Power Source		Calculated			DC Energy Meter	
dc volt	dc current	Power	Time (Min)	Energy(wh)	Energy Reading	Accuracy
12	5	60	5	300	301.526	0.51
24	5	120	5	600	599.328	-0.11
48	5	240	5	1200	1198.248	-0.15
96	5	480	5	2400	2392.873	-0.30
156	5	780	5	3900	3892.981	-0.18
192	5	960	5	4800	4779.87	-0.42
252	5	1260	5	6300	6277.942	-0.35
276	5	1380	5	6900	6862.547	-0.54
300	5	1500	5	7500	7468.739	-0.42

Table 2. Energy Accuracy at 15A

DC Power Source		Calculated			DC Energy Meter	
dc volt	dc current	Power	Time (Min)	Energy(wh)	Energy Reading	Accuracy
12	15	180	5	900	902.658	0.30
24	15	360	5	1800	1795.737	-0.24
48	15	720	5	3600	3620.653	0.57
96	15	1440	5	7200	7150.352	-0.69
156	15	2340	5	11700	11730.652	0.26
192	15	2880	5	14400	14350.916	-0.34
252	15	3780	5	18900	18845.725	-0.30
276	15	4140	5	20700	20782.529	0.40
300	15	4500	5	22500	22698.236	0.88

Above experiment was done with different voltage and different current and accuracy was measured

4 Conclusion

In this paper, the methodology for development of dc energy meter for dc micro grid is discussed. The development of indigenous test system will be beneficial for Indian industry at low cost. The best suitable method for this application i.e. sensing of voltage, current, DC energy calculation, temperature, humidity & pressure sensing is presented in this paper, which is cost effective too. The programming for Arduno Uno controller is done and it works as per the application. GUI software is developed in visual basic and it is vary user friendly. This GUI software provides weather data monitoring as well as dc energy data storing with real data and time parameter. Moreover, this system can be design if we need IoT based implementation. For that we will use UART to web server interfacing via WIFI module due to the need of smart city project & protect the data theft.

References

1. Development of Ardunio based IoT metering system for on demand energy monitoring by K.C.Okafor, G. C. Ononiwu in international journal of mechatronics
2. Design of a Smart Energy Meter with Overload Trip Facilities by Naeem Javed, Muhammad Aamir Shafi, Nasrullah Khan, Mohsin Khan in conference of international journal of engineering and advanced technology (IJEAT)
3. https://patents.google.com/patent/US4823075A/en?q=hall+effect&q=sensor&q=voltage&q=sensor&oq=hall+ef
4. https://patents.google.com/?q=iot&q=based&q=energy&q=meter&oq=iot+based+energy+met

A time-of-use based electricity price package mechanism in retail market considering consumers' diversity

Gang Luo[1], *Yujun* He[2,*], *Chen* Zhao[1], *Xuan* Zhang[1], *Shaohua* Lin[1], *Yue* Zhao[1], and *Yuanji* Cai[2]

[1]Power Dispatch and Control Centre, Guangdong Power Grid Co. Ltd., Guangzhou 510000, China
[2]State Key Lab of Control and Simulation of Power Systems and Generation Equipments, Dept. of Electrical Engineering, Tsinghua University, Beijing 100084, China

Abstract. In the deregulated retail market, consumers should have more right of choice for paying for their energy demand. This paper focuses on the design of electricity price package mechanism in retail market considering the demand elasticity of consumers. An optimal price package mechanism is proposed to incentivize consumers for peak-clipping/valley-filling. Consumers are able to choose the appropriate one from a set of price packages, each of which consists of specific time-of-use energy price and maximum-demand price. A numerical case study has shown the usefulness and effectiveness of the proposed mechanism.

1 Introduction

With the worldwide reform of power retail market deregulation, the consumers are becoming more active in the power trading, which leads to a more flexible power system where the demand can be adjusted to follow with the variation of generation. Instead of demand-side management (DSM) by administrative means, nowadays the demand response (DR) by means of dynamic price or incentives is playing an important role to balancing the demand/generation and optimally allocate power system resources [1]. In this context, the utility companies or retailers have to design more appropriate pricing models to attract consumers in the market competition [2].

Many studies have investigated the design of pricing mechanism in retail market. The dynamic price mechanisms, including time-of-use (TOU) pricing and real time pricing (RTP), are proved to solve the short-term lack of system capacity and improve the economic surplus [3]. The demand elasticity [4-5] and load factors [6-7] at demand side are usually considered by utility companies in designing the retail price. On the basis of demand-price elasticity theory, the work in [8] proposed an optimal retail price model considering the distribution network constraint. In [9], the price assessment method about the user's satisfaction is discussed in terms of energy consumption behaviour of consumers and electricity bills. In [10], a framework for consumer's power demand based on TOU contract characteristics is proposed for minimizing the mean electricity price paid by the customer. In [11], a dynamic multi-objective optimization of TOU price under multi-model structure is investigated. However, the price package mechanism based on TOU price is rarely studied considering detailed models of consumers' behaviour.

This paper focuses on designing an appropriate price package mechanism considering consumers' diversity. Both the load factors and demand elasticity functions are considered to establish a detailed model to describe the behaviour of different consumers with the given price packages. A two-settlement framework for each price package is adopted, including TOU energy price and capacity price related to maximum demand of consumer. The optimization is formulated considering the behaviour of different consumers with the proposed price packages. The rest part of this paper is organized as follows: Section 2 presents the clustering approach for consumers in retail market. The TOU based electricity package mechanism are introduced in Section 3. Then, the optimization of electricity package is formulated in Section 4 and a numerical case study is conducted in Section 5. Finally, Section 6 gives brief conclusions about the proposed mechanism.

2 Clustering approach for consumers in retail market

2.1 Classification of consumers

Electricity consumers can be classified in different manners based on their position, size or nature of use. The utility company usually apply different tariffs for the different consumers. In this paper, we classify the consumers using their patterns of consumption, including load factors and demand elasticity factors. Specifically, data dimension reduction technology is adopted for a context of TOU with periods of peak hours (PH), off-peak hours (OH) and valley hours (VH).

* Corresponding author: yujunhe@mail.tsinghua.edu.cn

2.1.1 Load factors

Since the load curve of consumers is too complicated for the clustering analysis, a data dimension treatment is needed. The dimension of load curve can be reduced by adopting a series of load factor parameters based on TOU. The load factor parameters are shown in Table. 1.

Table 1. Parameters in the optimization (m=monthly)

Parameter	Math. Expression
Daily load factor	$\phi_i = P_{av,i} / P_{\max,i}$
Load factor of PH	$\phi_i^{PH} = P_{av,i}^{PH} / P_{av,i}$
Load factor of OH	$\phi_i^{OH} = P_{av,i}^{OH} / P_{av,i}$
Load factor of VH	$\phi_i^{VH} = P_{av,i}^{VH} / P_{av,i}$

2.1.1 Demand elasticity

The demand elasticity reflects the responsiveness of consumers to the variation of price. These patterns depend on the load flexibility, users' habit and their willingness to provide demand response. The demand elasticity factors of consumers can be expressed as the ratio of load variation divided by the ratio of price variation during the intra-hour or inter-hours, respectively shown in (1) and (2).

$$E_{i,t\text{-}t} = \frac{\Delta Q_{i,t}}{\Delta \lambda_t} \cdot \frac{\lambda_t}{Q_{i,t}} \quad t \in T \tag{1}$$

$$E_{i,t\text{-}t'} = \frac{\Delta Q_{i,t}}{\Delta \lambda_{t'}} \cdot \frac{\lambda_{t'}}{Q_{i,t}} \quad t,t' \in T \& t' \neq t \tag{2}$$

where $E_{i,t\text{-}t}$ is the intra-hour demand elasticity factor at hour t of consumer *i*, $E_{i,t\text{-}t'}$ is the inter-hour demand elasticity factor at hour *t* in response to the price variation at hour *t'*. λ_t and $\Delta\lambda_t$ are respectively the initial price at t hour and its variation, $Q_{i,t}$ and $\Delta Q_{i,t}$ are respectively the initial demand and its variation at hour t.

2.2 Clustering of consumers

In this work, we adopt K-means method to for clustering consumers. K-means clustering aims to partition n observations into k clusters in which each observation belongs to the cluster with the nearest mean, serving as a prototype of the cluster. Given the vector of consumer i is $\mathbf{x}_i = \left[\phi_i, \phi_i^{PH}, \phi_i^{OH}, \phi_i^{VH}, E_{i,xH\text{-}xH}\right]$ and a set of consumers $(\mathbf{x}_1, \mathbf{x}_2, \ldots \mathbf{x}_i, \ldots, \mathbf{x}_N)$, The K-means method will compute I (≤ N) sets S = {S1, S2, ..., SI} so as to minimize the within-cluster sum of squares (WCSS). A standard K-means algorithm is based on the iterative process: the first step is to assign each consumer to the cluster whose mean has the least squared Euclidean distance $\|\mathbf{x}_i - m_p\|^2$, where m_p is the centroid of the cluster *p*; then m_p is updated using the consumers in the cluster.

$$m_p = \frac{1}{|S_p|} \sum_{x_i \in S_p} x_i \tag{3}$$

3 TOU-based electricity price package mechanism

The TOU pricing mechanism is a typical time-varying tariff differentiated during different periods. TOU prices encourage consumers to shift the use of electricity towards low-demand periods so that the capacity infrastructures in the power system can be fully used by reducing the power gap between peak and valley. The TOU is a sort of differentially pricing method for time domain, whereas it is identical for all consumers. The TOU prices are shown in (4), where λ_{PH}, λ_{OH} and λ_{VH} are respectively the price during PH, OH and VH.

$$\lambda_t^{TOU} = \begin{cases} \lambda_{PH} & t \in PH \\ \lambda_{OH} & t \in OH \\ \lambda_{VH} & t \in VH \end{cases} \tag{4}$$

Similarly, the differentially pricing for different users is called electricity price package. By offering different electricity price packages in the retail market, consumers are possible to choose one package that benefits themselves in terms of maximizing their own economic surplus. Besides, this approach can improve the extent of liberalization of the retail market.

A two-settlement pricing mechanism with energy and capacity prices is widely adopted in China's power retail market. In this paper, we propose a TOU based electricity price package design, each of which consists of specific TOU prices and a maximum-demand price to pay for the capacity. Let λ_j^{cap} be the maximum-demand price and λ_j^{PH}, λ_j^{OH}, λ_j^{VH} be the TOU prices of price package j during PH, OH and VH. The demand consumption amount of typical consumer *i* during PH, OH and VH are respectively $Q_{i,j}^{xH} = D_i^{xH}(\lambda_j^{xH})$, *xH=PH, OH, VH*. Supposing the monthly demand amount of consumer *i* is $Q_{i,0}$. Thus, $Q_{i,j}^{xH}$ can be expressed as:

$$Q_{i,j}^{xH} = Q_{i,0}^{xH} \cdot \left(\begin{array}{c} k_i^{xH} + \frac{E_{i,xH\text{-}PH}}{\lambda_0^e} \lambda_j^{PH} + \\ \frac{E_{i,xH\text{-}OH}}{\lambda_0^e} \lambda_j^{OH} + \frac{E_{i,xH\text{-}VH}}{\lambda_0^e} \lambda_j^{VH} \end{array} \right) \quad (5)$$

where $k_i^{xH} = 1 - E_{i,xH\text{-}PH} - E_{i,xH\text{-}OH} - E_{i,xH\text{-}VH}$, $Q_{i,0}^{xH}$ is the part of $Q_{i,0}$ during xH, λ_0^e is the baseline energy price.

Let T_{xH} be the total hours of peak/off-peak/valley periods, $Q_{i,0}^{xH}$ can be estimated in (6):

$$Q_{i,0}^{xH} = \frac{T_{xH} \cdot \phi_i^{xH}}{24h \cdot \phi_i} Q_{i,0} \quad (6)$$

where ϕ_i is the daily load factor of consumer i, ϕ_i^{xH} is the load factor during xH, $xH=PH, OH, VH$.

The maximum demand $P_{i,j}^{\max}$ of consumer i when he chooses the price package j is expressed in (7).

$$P_{i,j}^{\max} = \frac{Q_{i,j}^{PH} + Q_{i,j}^{OH} + Q_{i,j}^{VH}}{24h \cdot \phi_i} \quad (7)$$

4 Optimization formulation for TOU-based price package

The utility company hopes to maximize its revenue from consumers when designing the price packages. Supposing the total number of consumers is N, the number of clusters of consumers is I, each consumer is represented by the typical consumer i which is the central element of the cluster that the consumer belongs to. The overall revenue of the utility company is expressed as:

$$B_{UC}^{total} = \sum_{i=1}^{I} \omega_i N \sum_{j=1}^{K} p_i^j B_i^j \quad (8)$$

where ω_i is the percentage of typical consumer i over N consumers. p_i^j is the probability that consumer i chooses the price package j, B_i^j is the payment of consumer i with package j expressed as:

$$B_i^j = \sum_{\substack{xH=PH,\\ OH,VH}} \lambda_j^{xH} \cdot D_i^{xH}(\lambda_j^{xH}) + P_{i,j}^{\max} \lambda_j^{cap} \quad (9)$$

The overall demand of N consumers during xH is:

$$Q_{total}^{xH} = \sum_{i=1}^{I} \omega_i N \sum_{j=1}^{K} p_i^j D_i^{xH}(\lambda_j^{xH}), \quad xH \in \{PH, OH, VH\} \quad (10)$$

The energy costs during different periods are impacted by the diversity of power generation. Without loss of generality, we suppose the energy cost during different cost is in a quadratic form, and the operating cost of each price package is ρ, the overall cost of the utility company is expressed as:

$$C_{UC}^{total} = \sum_{\substack{xH=PH,\\ OH,VH}} \left[\begin{array}{c} a^{xH} \left(Q_{total}^{xH}\right)^2 + \\ b^{xH} Q_{total}^{xH} + c^{xH} \end{array} \right] + \rho K \quad (11)$$

where a^{xH}, b^{xH} and c^{xH} are cost coefficients of xH period.

It is worth noting that, as the utility company is supposed to provide public service, it should not rise the total energy bill for consumers. Let λ_0^e be the baseline energy price for consumers and λ_0^{cap} be the baseline maximum-demand price (we assume all consumers are with unique price). The total revenue of the utility company without price package design can be obtained in (12).

$$B_{UC}^{non} = \sum_{i=1}^{I} \omega_i N \cdot \left[\begin{array}{c} \lambda_0^e \sum\limits_{\substack{xH=PH,\\ OH,VH}} D_i^t(\lambda_0^e) + \\ \lambda_0^{cap} P_i^{\max}(\lambda_0^e) \end{array} \right] \quad (12)$$

Therefore, we can formulate the optimization problem in which the prices of each package are determined for the objective of maximizing the net profit of utility company while the cost of consumers is not added, as shown in (13):

$$\begin{aligned} & \max_{\lambda_j^{cap}, \lambda_j^{xH}} f_{UC}^{total} = B_{UC}^{total} - C_{UC}^{total} \\ & s.t.\ B_{UC}^{total} \le B_{UC}^{non}, \\ & \lambda_{\min}^e \le \lambda_j^{xH} \le \lambda_{\max}^e, \\ & \lambda_{\min}^{cap} \le \lambda_j^{cap} \le \lambda_{\max}^{cap}, \\ & \forall j \in [1, K], xH = \{PH, OH, VH\} \end{aligned} \quad (13)$$

where $\lambda_{\min}^e$ and $\lambda_{\max}^e$ are the lower/upper bounds of energy price, $\lambda_{\min}^{cap}$ and $\lambda_{\max}^{cap}$ are the bounds of maximum-demand price. The hyper-parameter K can be adjusted outside of this optimization.

The entire algorithm of the price package design is shown in Fig 1. Firstly, pre-treatment of consumer data is performed by reducing the data dimension and identifying specific TOU parameters of each consumer; consumers are clustered using K-mean method to obtain

 https://doi.org/10.1051/e3sconf/202018202009

I-number typical consumers; then, we initialize the optimization conditions and let K=1, by computing the optimization and comparing the net profit of utility company with K packages and that with K-1 packages, we can find the best K and the optimal prices in the packages.

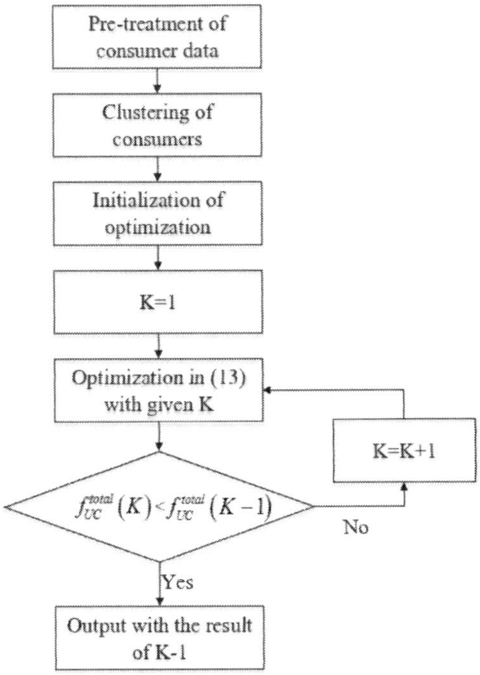

Fig. 1. Algorithm of the price package design.

5 Case study

For identifying the proposed method, a numerical case study is conducted in a real scenario with 100 residential consumers. The load curves of consumers are extracted from a database of Guangdong Grid in Southern China. The clustering results of the 100 consumers can be represented by three typical consumers including peak-demand consumer, smoothly-shaped consumer and anti-peak consumer, as detailed in follows:

1) Peak-demand consumers, who tend to consume more energy during peak hours even though the price is higher;

2) Smoothly-shaped consumers, who have a smooth load curve with high load factor;

3) Anti-peak consumers, who are willing to fill the valley demand if the price is lower.

The peak-valley TOU periods are set as: 14h00~17h00 and 19h00~22h00 for PH, 0h00~8h00 for VH, and other hours for OH. The costs parameters can be found in [12]. Other parameters for the optimization are shown in Table 2, some of which are based on the Guangdong power market practices.

The optimization is performed using CPLEX12.1 toolbox installed in Matlab r2016b. K is varied from 0 to 4, in which K= 0 is the baseline case without price package and prices are not changed, K=1 is actually the case of TOU pricing without choice for consumer, K=2~4 means there are some packages optional for consumers. The obtained results are shown in Fig. 2 and Fig. 3. In Fig. 1, the cost of utility company can be reduced when adopting price packages. Fig. 3 shows the net gain from price packages compared with the case K = 0. It can be observed that the result with K=3 price packages in this scenario is best. Fig. 4 shows the contribution of price packages in reducing peak demand and filling valley demand. When K=2, the peak demand is reduced at the most, then when K increases, the reduced peak demand is decreased. This is because the price packages give consumers more options if K is greater. The optimal price packages with K=3 are shown in Table 3.

Table 2. Parameters in the optimization (m=monthly)

Parameter	Value
λ_0^e	0.61￥/kWh
λ_0^{cap}	32￥/kWh/month
$\lambda_{\min}^e$	0.10￥/kWh
$\lambda_{\max}^e$	1.2￥/kWh
$\lambda_{\min}^{cap}$	30￥/kWh/month
$\lambda_{\max}^{cap}$	35￥/kWh/month
ρ	1000￥/month

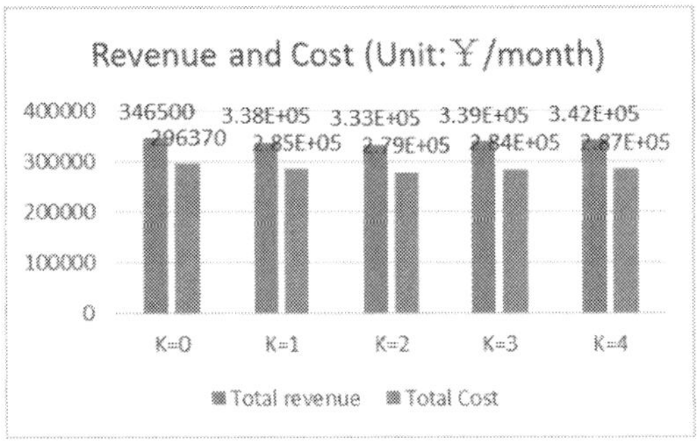

Fig. 2. Revenue and cost with different number of price packages.

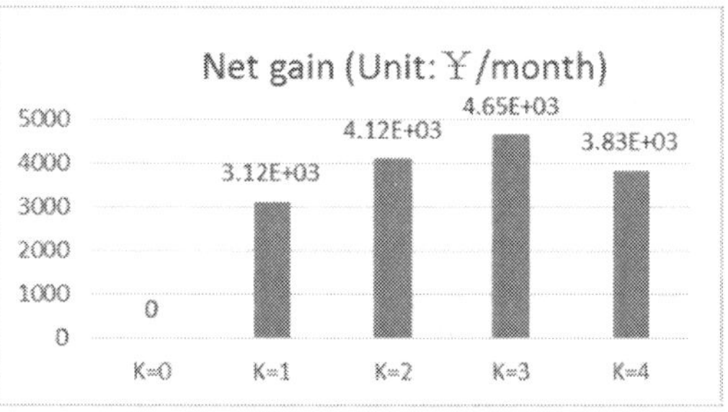

Fig. 3. Net gain with different number of price packages.

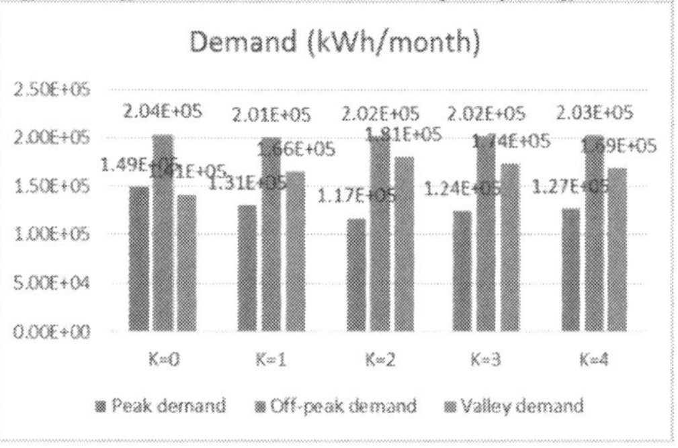

Fig. 4. Demand during different periods with price packages.

Table 3. Optimal price package results when K=3

Price package No.	1	2	3
λ_j^{PH} (￥/kWh)	1.05	0.84	0.7
λ_j^{OH} (￥/kWh)	0.61	0.64	0.61
λ_j^{VH} (￥/kWh)	0.20	0.43	0.55
λ_j^{cap} (￥/kWh/month)	35	32	30

6 Conclusions

This paper has investigated a novel TOU price package design mechanism in retail market. The price parameters in the packages are optimized considering the diversity of consumers in terms of load factors and demand elasticity coefficients. The numerical results have shown the effectiveness of the proposed method. With price packages (K is greater than 1), the profit of the utility company can be augmented while the consumers are endowed with more right of choice in retail market. Moreover, detailed model with multiple retailers in a competitive environment could be studied in the future work.

This work was supported by Guangdong Power Grid Co., Ltd. Project (No. 036000KK52180031).

References

1. Y. Bin, S. Kun, L. Dezhi, *et al.*, Research on Power Flexible Load Regulation Technology Based on Demand Response, in proceedings of *2018 8th International Conference on Electronics Information and Emergency Communication (ICEIEC)*, Beijing, pp. 150-153, (2018)
2. YAO D., CHU Y., Pricing Strategy of Electricity Price Packages under the Power Sale Side Reform in China, *Power & Energy*, 39(**01**): pp.59-61, (2018)
3. R. Li, Z. Wang, S. Le Blond and F. Li, Development of time-of-use price by clustering techniques, in proceedings of *2014 IEEE PES General Meeting*, National Harbor, MD, pp. 1-5, (2014)
4. R. Fonteijn, M. Babar and I. G. Kamphuis, An assessment of the influence of demand response on demand elasticity in electricity retail market, in proceedings of *2015 50th International Universities Power Engineering Conference* (UPEC), Stoke on Trent, pp. 1-6, (2015)
5. M. Babar, P. H. Nguyen, V. Cuk *et al.*, The development of demand elasticity model for demand response in the retail market environment, in proceedings of *2015 IEEE PowerTech*, Eindhoven, pp. 1-6, (2015)
6. S. R. Konda, L. K. Panwar, B. K. Panigrahi *et al.*, Investigating the Impact of Load Profile Attributes on Demand Response Exchange, in *IEEE Transactions on Industrial Informatics*, vol. **14**, no. 4, pp. 1382-1391, (2018)
7. A. B. Arani, R. Yousefian, P. Khajavi *et al.*, Load curve charactristics improvement by means of optimal utilization of demand response programs, in proceedings of *2011 10th International Conference on Environment and Electrical Engineering*, Rome, pp. 1-5, (2011)
8. K. Feng, H. Zhou, Z. Liu and D. Hu, Retail market pricing design in smart distribution networks considering wholesale market price uncertainty, in proceedings of *IECON 2017 - 43rd Annual Conference of the IEEE Industrial Electronics Society*, Beijing, pp. 5968-5973, (2017)
9. HUANG Haixin, DENG Li, WEN Feng. Customer Response Behavior Based on Real-Time Pricing, in *Electric Power Construction*, **37** (2): 63-68, (2016)
10. Y. Hung and G. Michailidis, Modeling and Optimization of Time-of-Use Electricity Pricing Systems, in *IEEE Transactions on Smart Grid*, vol. **10**, no. 4, pp. 4116-4127, (2019)
11. Y. Hu, Y. Li and L. Chen, Multi-Objective Optimization of Time-of-Use Price for Tertiary Industry Based on Generalized Seasonal Multi-Model Structure, in *IEEE Access*, vol. **7**, pp. 89234-89244, (2019)

12. X. Zhang, W. Liu, Y. Chen, *et al.*, Electricity Market Design and Operation in Guangdong Power, in proceedings of *15th International Conference on the European Energy Market*, Lodz, pp. 1-5, (2018)

Fault Investigation and Operation Strategy of Dry Type Air-core Reactor in Severe Cold Environment

Lin Wang[1,*], *Jianquan* Liang[2], *Wenguang* Hao[3], and *Hongyu* Wang[3]

[1]China Electric Power Research Institite, High Voltage Department, No. 15 Qinghexiaoying Road, Beijing, China
[2] Electric Power Research Institute State Grid Heilongjiang Electric Power company Limited, No.7 Xiangjiang Road, Harbin, China
[3]Beijing Power Equipment Group Co.Ltd, No. 12 Haotian Road, Beijing, China

Abstract: The influence of extreme climate such as severe cold temperature on power transmission system can not be ignored, the low temperature could cause complicated mechanical and dielectric deterioration to electrical materials and devices. In China's recent operation experience of reactors, it is found the burning fault rate of dry type air-core reactors in cold northeast region was obviously higher than that of other regions, which had inflicted negative impact on substation equipment and personnel safety. To tackle with the problems and prevent potential faults, typical fault characteristics and mechanism at low temperature, the specific influence of low temperature on materials and reactor performance, as well as corresponding operation strategies would be studied and proposed in this paper.

1 Introduction

At extreme low temperature, transmission system faults may occur such as wire icing, metal fittings breaking, deformation and cracking of outer porcelain and synthetic insulation materials, and also circuit breaker locking caused by SF_6 pressure drop [1-4]. Since the extreme low temperature could cause complicated mechanical and dielectric deterioration to electrical materials and devices, the fault characteristics and operation strategy of electrical devices at low temperature have been paid more attention to enhance operation quality and safety.

In China, the northeast region is very cold in winter, the average temperature is about -20 ℃ , some areas could even reach -40℃ [5], in such a harsh operating environment, power transmission system and related equipment are facing great challenges. In recent years, China's power transmission projects have been developed in a more compact trend, with less floor space for transmission lines and larger transmission capacity, thus leading to increasing demand for reactors for reactive power compensation and filtering functions [6]. However, in recent operation experience of new commissioned reactors, it is found that compared with other regions in China, the burning fault rate of dry type air-core reactors in cold northeast region was obviously higher, which has inflicted negative impact on substation equipment and personnel safety [7-10].

In order to improve reactor operation safety and quality, typical low temperature fault characteristics were summarized based on fault investigation, the fault mechanism and low temperature impact on equipment were analysed by simulation and material test, and strategies aimed at reactors operation at extreme cold temperature were proposed in this paper.

2 Fault investigation

2.1 Dry type air-core reactor structure

The main structure of the dry type air-core reactor is composed of several concentric cylindrical encapsulations, which are parallel connected. In each encapsulation, there are several shunted coils (commonly aluminium) with high dielectric strength of insulation coating, functioning as turn-to-turn insulation. The encapsulation of dry type air-core reactors is wrapped with glass fibre material impregnated with epoxy resin, and a glass wire drawing rod is used as strut between encapsulations to form a heat dissipation port, as shown in figure 1.

The coil ends on the top and bottom are welded on aluminium frames, which act as electrical connection also mechanical compression. The whole structure is heat curing sandblasting treated, and a layer of special organic silicon is coated to resist ultraviolet radiation.

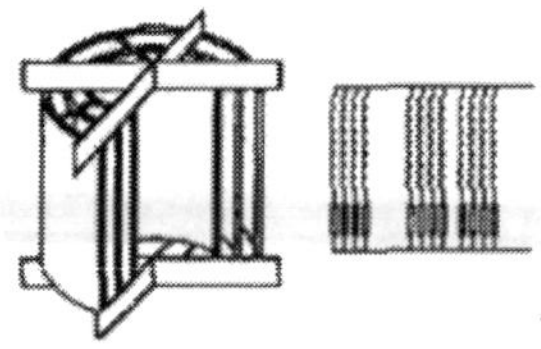

Fig. 1. Structure of dry type air-core reactor

* Corresponding author: wanglin@epri.sgcc.com.cn

2.2 Fault investigation and analysis

12 recent typical burning failures of 66 kV dry type air-core shunt reactor in a northeast province of China were investigated and analysed, findings as below:

- 10 of 12 faults occurred in winter, accounting for 83% of the total.
- Except for 1 fault that caused by bad contact of turning-regulating ring, the other faults were all caused by inter-turn short circuit.
- Most of the discharge traces appear in the outer encapsulation layers.
- Surface cracks were found on most fault reactors, while impurities such as water stains and bird droppings were found.
- After some fault reactors were disintegrated back in the factory, it was found that manufacture defect like burrs, bubbles and uneven distribution of insulation materials appeared in inter-turn insulation of the reactors.

Since cracks and impurities were found in most of the fault reactors, the specific fault morphology in the case of cracks and bird droppings was described as follows:

After disassembly of some fault reactors, many small creepages were found on the surface of reactors, most of which occurred in the transverse cracks near the strut and lead exit. The initial creepages developed horizontally, and the more serious creepages began to develop longitudinally along the strut. Meanwhile, bird droppings were usually found near the discharge marks, the traces of discharge were illustrated as figure 2:

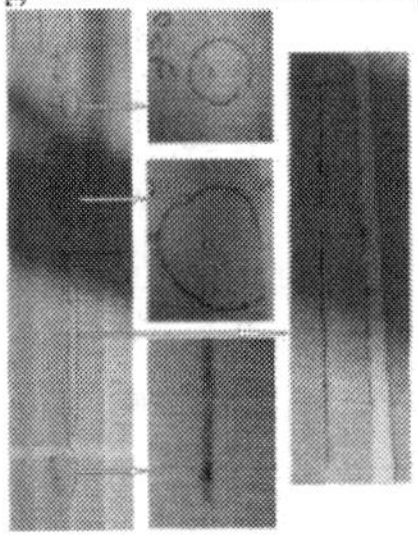

Fig. 2. Reactor discharge with cracks and bird droppings

From the discharge picture, it could be found that discharge developed along bird droppings, and mostly located near the lead exit and strut, and most of the starting points of discharge were accompanied by obvious cracks. For those cracks located far from electric field concentration area, no obvious discharge marks were found.

According to the fault investigation result above, it can be inferred that the typical process of dry-type air-core reactor burning failure at extreme low temperature is that: unbalanced stress release of reactor materials caused the form of transverse cracks on the insulation surface, and cracks near the electric field concentration area such as lead exit strut might lead to more serious electric field distortion. Besides, impurities such as water and bird droppings could enter into inner structure of reactor through the cracking parts, resulting in the insulation deterioration. The insulation weakness and distorted electric field both contributed to discharge and inter-turn short circuit, when inter-turn short circuit happened, loss of reactor could increase sharply, then followed rapid burning. In addition, if the reactor itself had material or manufacturing defects, such as burrs and bubbles in inter-turn insulation or wires that lack of tensile strength, the possibility of insulation deterioration and inter-turn short circuit fault would further increase.

3 Fault mechanism analysis

In the above chapter, the process and possible influencing factors of typical reactor burning faults at extreme low temperature were analysed based on fault investigation. Further analysis of the fault mechanism, and the specific influence of low temperature on reactor property were discussed in the following:

3.1 Cracking mechanism of encapsulation

The main wire material of dry type air-core reactor is aluminium, coated with polyester film, and the insulation between strands is also polyester film. In order to ensure insulation and isolation from the external environment, each encapsulation is composed of multiple layers of shunted wire, coated with fiberglass and epoxy composite. Due to the difference in thermal expansion coefficient between the wires and encapsulation insulation material (aluminium 2.3×10^{-5}/K, epoxy/fiberglass composite 2.7×10^{-6}/K), thus when the temperature changes, the original solidly combined structure of wire and insulation will shrink in different amplitude, then inflict stress on the joint part of two materials. If the tensile/compressive strength of epoxy glass and wire could not withstand the stress produced by unbalanced shrinkage and electric stress produced by short circuit current during operation, the material and reactor structure deformation happens, and further develop into surface cracking.

Therefore, the deformation and cracks of reactor is determined by not only the actual stress produced by shrinkage and short circuit current, but also the mechanical tensile, compressive, bending strength of the materials.

3.2 Material mechanical strength analysis

In order to study the mechanical strength of reactor windings and insulation material, the epoxy resin and aluminium materials were bonded together by adhesive and then divided into small pieces to simulate wire with insulation coating. Then tensile tests for those aluminium/ epoxy resin pieces were carried out under different temperatures. The test pieces were stretched with gradually increased mechanical load, until the aluminium and epoxy resin were completely detached from each other. The tests were conducted at room temperature 27 ℃, -25 ℃ and -40 ℃, respectively, and repeated several times at each temperature. The withstand load for the samples were shown in table 1:

Table 1. Tensile test result of epoxy wire bonding

Sample number	27 ℃ (kPa)	-25 ℃(kPa)	-40 ℃(kPa)
1	340.3	322.8	311.8
2	349.5	259.4	413.9
3	339.4	255.7	264.0
4	392.7	372.5	329.3
5	341.2	417.5	248.3
6	342.1	428.6	249.2
7	456.2	262.1	240

It could be seen that the low temperature environment had a great influence on the maximum withstand load of the samples, the tensile strength decreased obviously with the decrease of temperature. The average maximum withstand load is 365.9 kPa at room temperature of 27 ℃, 331.2 kPa at -25 ℃, which is 10% lower than at room temperature, the withstand load fell to only 293.8 kPa at -40 ℃, which is 20% lower than at room temperature.

This test result may be caused by the different degree of shrinkage of winding wires and epoxy resins at low temperature, and may also be related to property changes of adhesives. The decrease of the maximum withstand load indicated that the bonding performance between two materials were weakened compared with that at normal temperature, which makes the joint surface more fragile. When the reactor materials undertook transient electric force or unbalanced stress, the cracks between insulations and windings were more easily to form due to the declined mechanical strength at low temperature.

3.3 Winding force analysis

During operation, the dry type air-core reactors could suffer mechanical stress such as tensile, vibration and extension force under alternating magnetic field, with the accumulation of these stresses, deformation or cracks of the reactor structure might happen. The stress of the reactor windings and influence of low temperature on the stress were analysed by simulation as follows:

The reactor model composed of three layers of windings was established, and epoxy strut between layers was set. The winding force was simulated with current RMS of 300A, at temperature of 27 ℃, -25 ℃ and -40 ℃. In order to fully consider the current distribution in the winding, field-circuit coupled method was used, the current source with amplitude of 424.26 A, frequency of 50 Hz was the input.

The magnetic field force density vector diagram of the reactor winding was shown in figure 3. It could be seen that the overall force on each winding was outward expanded under the interaction of current and magnetic field.

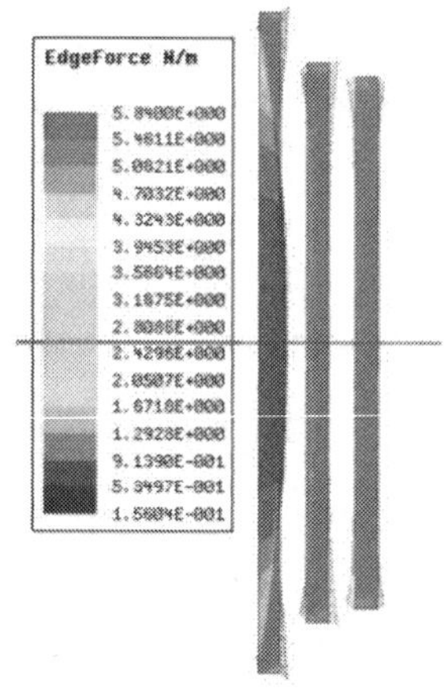

Fig. 3. Force density of reactor winding

Besides, a 3D model of reactor structure was established to study the stress and strain of the reactor windings under the action of the electromagnetic force. The windings, encapsulation and strut structures were all fully considered in the model. The axial view and top view of the model were shown in figure 4.

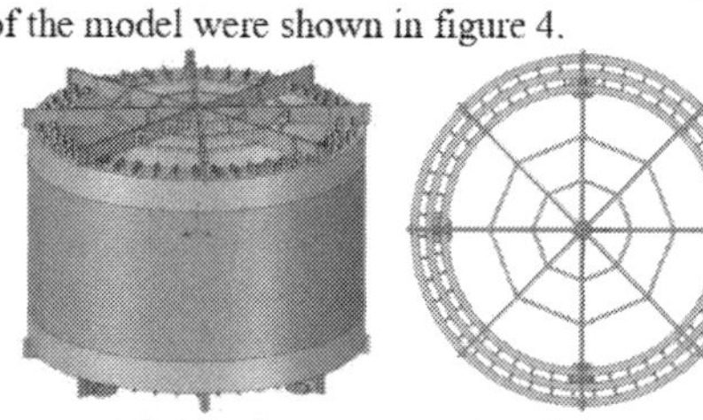

axial side view top view

Fig. 4. 3D model of reactor structure

After subdivision of the model and establishment of a structure and electromagnetic coupled simulation environment, the electromagnetic force results were mapped to the structure model to simulate material strains. Figure 5 showed the electromagnetic density vector and deformation diagram. Due to the overall outward expansion, the lifting boom was concaved to the inside, and the deformation of each winding layer of the reactor showed a trend of large deformation in the middle and small deformation at both ends.

Fig. 5. 3D Electromagnetic model and deformation

In order to study the influence of extreme low temperature on the stress of the reactor, the stress of the reactor windings at temperature of -40℃, -25℃ and 27℃ were simulated and analysed. The stress diagram at three temperatures were shown as figure 6 to 8:

Fig. 6. Reactor winding stress at -40℃

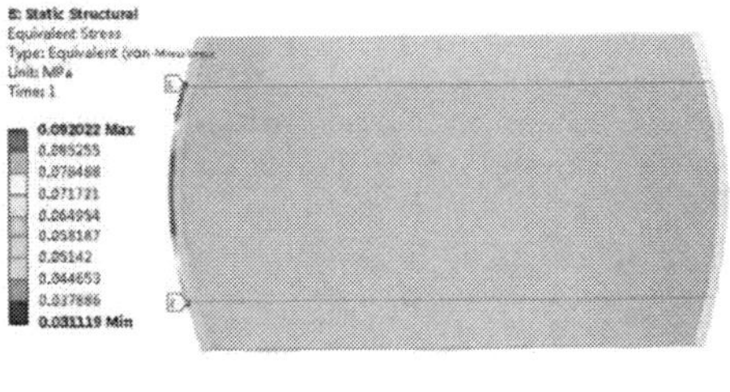

Fig. 7. Reactor winding stress at -25℃

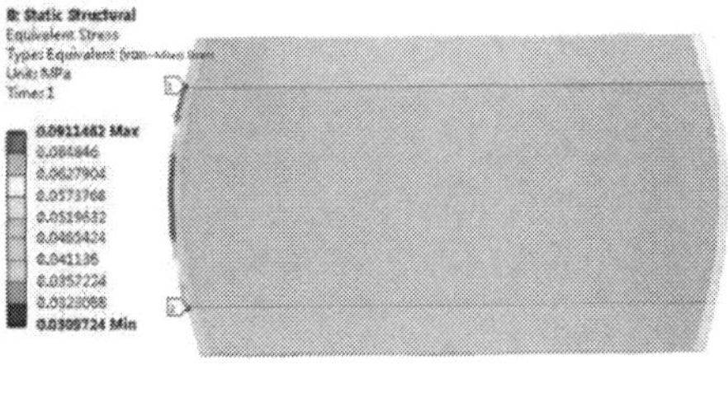

Fig. 8. Reactor winding stress at 27℃

According to simulation result above, the magnetic field intensity in the centre of the reactor was the largest while the outer magnetic field was relatively small. It is also found that when temperature decreases, the overall stress of the winding tends to increase. When the temperature drops to -40℃, the stress of the reactor increases by about 2% compared with the normal temperature. Though the stress difference was slight, it might impact more with the accumulation effect.

3.4 Fault mechanism at low temperature

Based on the study of material mechanical strength and reactor winding force, it can be concluded that at low temperature, the stress suffered by reactor slightly increased, while the mechanical strength of main reactor materials decreased. Thus, compared with moderate temperature, extreme low temperature could be more likely to facilitate insulation deformation and cracking.

Besides, according to fault investigation and disassembly evidence, the material and manufacturing defects would also be a main cause of the reactor fault. Therefore, the manufacturing defects and extreme cold environment were considered to be the two main reasons for the failure of dry air core reactor in winter. Detailed analysis of the two factors were discussed as follows.

3.4.1 Material and manufacture defect

According to the fault investigation, manufacturing defect were found in considerable faults cases. If material or manufacturing defect existed in reactor, For example, burr, bubble and uneven distribution of insulating materials in turn to existed in inter-turn insulation, it would lead to poor heat and stress resistance performance of the reactor, with high possibility of insulation breakdown. For example, there are burr, bubble and uneven distribution of insulating materials in turn to turn insulation, which are more likely to cause insulation damage and failure under stress. Besides, manufacturing defects such as coil solidification problems will accelerate the formation air gap and uneven material distribution.

External environment such as extreme cold could aggravate those material and manufacturing defects and develop into more serious faults. The shrinkage mismatch of different materials caused by low temperature could also aggravate the air gap and other defects. Small defects may develop into larger ones, which will lead to the declination of both mechanical and insulation properties of the whole structure.

3.4.2 External environment

Since the dry type air-core reactors are often operated outdoors, it could be affected by external environment to a great extent. At low temperature, the bonding performance and tensile strength of the windings were significantly lower than that of the normal temperature, making the windings more vulnerable. For the same force stress suffered, the reactor windings were more likely to deformed and brought about insulation damage at lower temperature.

In addition, in low temperature environment, the temperature gradient of the reactor is large, leading to different shrinkage of encapsulation insulation and wires materials. The expansion strain of the winding caused shear force on the encapsulation, leading to insulation damage and cracking. When the cracks formed, impurities such as humidity, bird droppings, dust would enter to further deteriorate the insulation and more prone to develop into severe fault.

4 Operation and maintenance strategy

According to the fault investigation and mechanism analysis, some corresponding operation and maintenance strategies were proposed to avoid typical fault.

4.1 Reactor inspection and cleaning

Routine inspection and cleaning of the upper end surface and air duct should be carried out. The internal air duct of the reactor could be selectively checked with an endoscope to observe whether there are foreign matters, dirt, or discharge traces. For the impurities such as dust in the air duct, it should be considered to use high-pressure wind for purging. For bird droppings and other traces in the air duct, it is recommended to use pure cotton cloth stained with alcohol to wipe them off.

4.2 Measures against bird droppings

Since the bird droppings will accelerate the electric field distortion and development of partial discharge, the access to the reactor needs to be protected. It is suggested to add bird proof device on the inner diameter of lower and upper part of the reactor to prevent birds from entering the reactor, also to cut off access of bird droppings and other impurities into the reactor.

4.3 Air duct protection measures

Some of the early reactor air ducts were not coated with anti-pollution flashover coating due to technical reasons. It is recommended to spray coatings on the inner air duct surface to block the way of discharge caused by moisture or contamination.

4.4 Annual inspection regulation

It is suggested that an annual test inspection regulation be set for all outdoor operated reactors. The test items including DC resistance, grounding and radial insulation performance test, on-site power frequency loss test, on-site turn to turn overvoltage test, and also arc extinguishing performance and service life of the switch should be checked. Besides, if conditions permit, on-line monitoring for reactors could be set to track the operation parameter.

5 Conclusion

• Based on operation fault investigation and analysis, the fault rate of dry type air-core reactors was higher at extreme low temperature, in which most reactor faults at low temperature were caused by inter-turn short circuit, usually with cracking and impurities found on the reactor surface;
• According to material test and reactor electromagnetic simulation, it was found that the stress reactor suffered slightly increased at low temperature, while the mechanical strength of main materials of reactors decreased. Thus, cold environment was considered to have contribution in promoting deformation and cracks forming;
• The promoted surface cracks forming and decreased mechanical strength would lead to electric field distortion, reactor structure deformation and even serious discharge accidents. Therefore, the dry-type air core reactor operated in the extremely cold environment should have higher electrical and mechanical strength.
• Since the thermal expansion coefficient of reactor winding is different from that of encapsulation, theoretically there would be considerable thermal stress inside the reactor structure at extreme low temperature. In addition, after the reactor is energized, the encapsulation would also be impacted by electric force of the winding, thus the stress distribution of internal reactor is very complex. For further study, the more accurate simulation and test analysis of internal stress of reactor, the comprehensive materials characteristics at low temperature should be continuously investigated, so as to guide reactor selection rule and operation method in extremely cold environment.

Supported by SGCC research project "Research on diagnosis technology for electric equipment in extreme cold environment" (GY71-17-056).

References

1. Zhou Y, Gao Z. Characteristics of wire icing and icing zoning in Hubei Province[J]. Journal of Natural Disasters, 2013, 22(6):169-177.
2. Yin, Quan. Experimental Study of Tensile Test of Line Fittings Steel under Low Temperature[J]. Advanced Materials Research, 2012, 619:541-544.
3. Maosheng C . Research on Mechanical Property and Critical Crack Size of Porcelain Post Insulator Under Low Temperature Environment[J]. Insulators and Surge Arresters, 2018.
4. Buret F , Beroual A . SF6 dielectric behaviour in a high voltage circuit breaker at low temperature under lightning impulses[J]. IEEE Transactions on Power Delivery, 1996, 11(1):267-273.
5. Qi J I . Characteristics of temperature and precipitation in Northeast China from 1951 to 2000[J]. Journal of Meteorology and Environment, 2006.
6. Pei-Hong Z , Hui-Wen H E, et al. Application of Controllable Reactors to 1000kV AC Compact Transmission Line[J]. high voltage engineering, 2011, 37(8):1832-1842.
7. LUO Tianming, NI Xuefeng, et al. Study on the Fault of Dry-type Air-core Reactor Under Low Temperature[J]. power capacitor & reactive power compensation, 2017.
8. WU Dongwen, Maintenance Branch. Temperature Field Distribution and Infrared Temperature Measurement Method Research of 35kV Dry-Type Reactor[J]. Transformer, 2013.
9. Jiang Z , Zhou H , Song J , et al. Temperature field calculation and experimental analysis of dry-type air-core reactor[J]. Transactions of China Electrotechnical Society, 2017.
10. Lai W, Luo H , Li W , et al. The research of the hot spot temperature of the dry-type reactor winding based on the inversed-heat conduction model[C]// International Conference on Materials Science. American Institute of Physics Conference Series, 2017.

The Reactive Power Support Strategy based on Dual-loop Control for Three-phase Grid-connected Inverter

WAN Qian[1,2*], *Xia* Chengjun[1], *Azeddine* Houari[2], *Zhao* Xue[1], *Xia* Chengjun[1], *Zheng* Xiaotian[1] and *Huang* Chuyin[1]

[1]South China University of Technology, College of Electric Power, Wushan Road, Guangzhou 510640, China
[2]IREENA Laboratory, University of Nantes, 37 boulevard de l'université, Saint-Nazaire 44602, France

Abstract. Renewable energy sources (RESs) generally connected with electric power system via power electronic interface. This paper presents a reactive power and voltage (Q/V) control strategy of three-phase photovoltaic (PV) system to offering reactive power based on the typical dual-loop control topology. It is worth mentioning that control strategy can support reactive power when a low voltage fault occurs in AC bus without additional compensation device. With the help of the decoupling control, the PV array can generate active power as much as possible in variable external solar radiation conditions. The voltage of PV arrays is adopted as the objective, which on account of the easy availability and controllability of voltage, to control output active power. Besides, accurately modeling process from a PV cell to PV array is described in the beginning to acquire the P-V and V-I characteristics of PV arrays, which promote the designment of Q/V control.

1 Introduction

Due to the conflict of the fast development of world economy and fossil energy lacking, and climate changes caused by power generation by burning of fossil fuels, it is inevitable to develop RESs on a large scale. Converting renewable energy like solar energy, wind power, geothermal energy, and bioenergy into electrical energy is the most efficient and proven technique so far. Electrical energy is an independent and interesting transient vector that provides a great way to incorporate RESs through power generations, power converters, transmissions, and energy storages.

Photovoltaic (PV) system is one of the most mature RESs power generation technologies in the world, which is also the geographically most distributed measures of electricity production worldwide. It makes full use of solar radiation energy with the advantages of modularization, scalability in the generation and simple installation. Besides, the solar radiation as clean, renewable energy, it produces no pollution and never run out, hence achieved the policy support. A photovoltaic cell firstly developed in Bell Laboratories, America. Until now, Japan and Germany have mastered the most advanced PV system technology in the world. And China has occupied the largest installed and total solar power capacity globally [1].

In the condition that PV arrays as the main electrical sources for supply power, it is worth mentioning that the inverter control strategy can support reactive power when a low voltage fault occurs in AC bus without additional compensation device. Therefore, this paper introduced the PV system in four parts. 1) the detailed and accurate modeling of PV arrays; 2) acquiring output characteristics of PV array, P-V and V-I parameters are included; 3) The typical dual loop control and the improved reactive power and voltage (Q/V) control for three-phase grid-connected inverter; 4) conclusions and discussions of simulation results.

2 Modeling of PV Array

A PV cell, or solar cell, is the most basic electrical unit that converters solar radiation into electrical energy. In normal conditions, multiple cells that connected in series or/and parallel constitute a module or panel, which played the role of protecting semiconductor wafer in cells with tempered glasses, encapsulant and so on [2]. In order to generate cumulative voltage or/and current, a PV array implemented by strings of panels that connected in parallel at the same time, with each string consists of modules connected in series.

2.1. Solar Cell Model

The equivalent circuit of a general PV cell is shown in Fig. 1. The practical PV module includes R_p, the parasitic parallel (shunt) resistance and Rs, the series resistance. Rs mainly due to the device itself that related with the semiconductor layer, while R_p is based on the leakage current of the *p-n* junction [3], [4].

Generally, series part offers very little resistance, and shunt resistance being very high. According to modeling requirements, the output current could be chosen either in practical PV model (which refer to I, as shown in Fig. 1) or in the ideal condition (I_{pv_ideal}) to represent the terminal

* Corresponding author: 614575573@qq.com

output of one PV module that depends on designers.

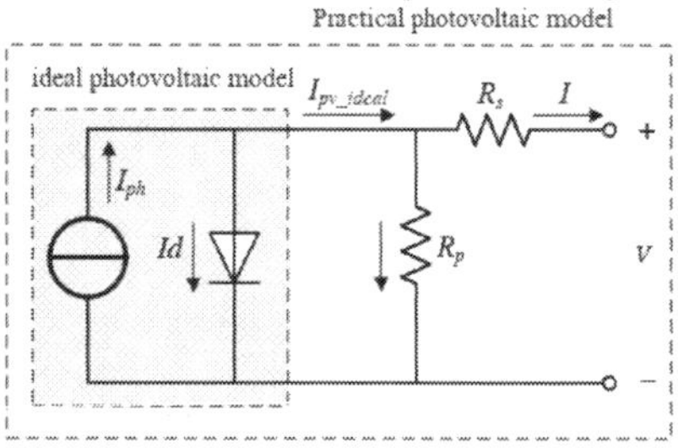

Fig. 1. A general equivalent circuit of diode PV cell model and the ideal PV mode.

The output current of a single PV module in both conditions can be written as:

$$I = I_{ph} - I_S\left(e^{\left(\frac{q(V+IR_s)}{AkT}\right)} - 1\right) - \frac{V + IR_s}{R_p} \quad (1)$$

Where I_{ph} is the light-generated current (or photocurrent) that describes relations between photocurrent and irradiations, temperature; I_S is the diode saturation current (or the cell saturation of dark current); q is an electron charge (1.602×10-19 C); k is the Boltzmann constant (1.3807×10-23 J/K); T is the working temperature of cell, A is an ideal factor (generally between 1 and 1.5, and mainly dependent on PV manufacture technology).

The light-generated current is mostly influenced by both solar radiations and working temperature, which is given as follows [4], [5]:

$$I_{ph} = \frac{G}{G_{ref}}\left(I_{ph,ref} + C_T\left(T - T_{ref}\right)\right) \quad (2)$$

Where $I_{ph,ref}$ is the light-generated current under standard conditions, generally refer to the external temperature at 25°C and irradiations is 1000W/m^2. In some cases, $I_{ph,ref}$ can be replaced by short-circuit current I_{SC} under standard conditions, which can be obtained from production datasheet [5]. G (W/m^2) is the irradiation on the device surface, and G_{ref} the solar radiation at the reference condition (1000W/m^2). T and T_{ref} (K) are the actual and reference temperatures respectively. C_T (A/K) is the manufacturer supplied temperature coefficient.

However, sometimes when R$_s$ is high enough and R$_p$ is very low, then cell model can be considered as an ideal device, then equation (1) can be rewritten as [5]:

$$I_{pv\_ideal} = I_{ph} - I_S\ e^{qV/AkT} - 1 \quad (3)$$

2.2 PV Array Model

To provide greater power, a PV array consists of a set of modules that are electrically connected in series or/and parallel circuit. This is the typical configuration for a PV array model to provide the required current and voltage. If cells that connected in arrays have the same characteristic parameters. Therefore, the equivalent circuit of the PV array that corresponds with the solar cell is shown in Fig. 2.

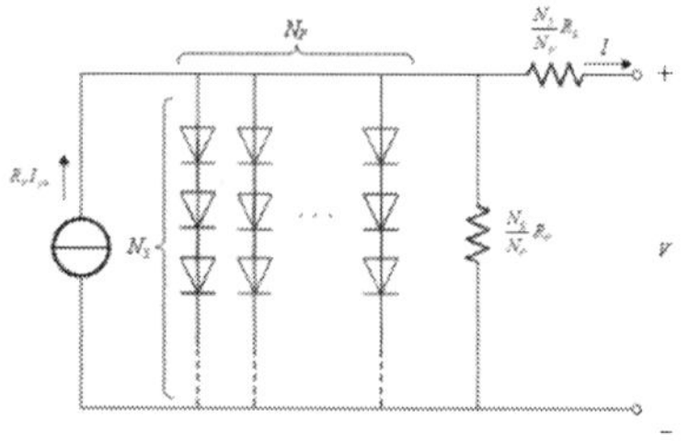

Fig. 2. Equivalent circuit for the PV module.

The equivalent circuit is shown in Fig. 2, the arrangement of cells in N_P parallel and N_S series respectively, the relation between voltage and current at the terminal of the circuit may be described as:

$$I = N_P I_{ph} - N_P I_S\left(e^{\frac{q}{AkT}\left(\frac{V}{N_S}+\frac{IR_s}{N_P}\right)} - 1\right) - \frac{N_P}{R_p}\left(\frac{V}{N_S} + \frac{IR_s}{N_P}\right) \quad (4)$$

Where E_g is material bandgap energy of the semiconductor in the PV cell (E_g = 1.12 eV for the polycrystalline silicon at 25°C). $I_{S,ref}$ is the reverse saturation current at the reference condition, it can be approximately written as the following relation:

$$I_{S,ref} = \frac{I_{ph,ref}}{e^{\left(qV_{OC}/N_s AkT\right)} - 1} \quad (5)$$

Where V_{oc} is the terminal voltage in open-circuit when current equal to zero. For the single solar cell, N$_S$ =1.

2.3 PV Characteristics

After the modeling of PV cells, its characteristics can be analyzed directly. The results of the Simulink described the output power, voltage, current and relations between them. For instance, relations between output power and voltage (P-V) when irradiations (G) set to 1000W/m^2 and 500 W/m^2 respectively are displayed in Fig. 3, with I-V (current and voltage relations) curves is also given.

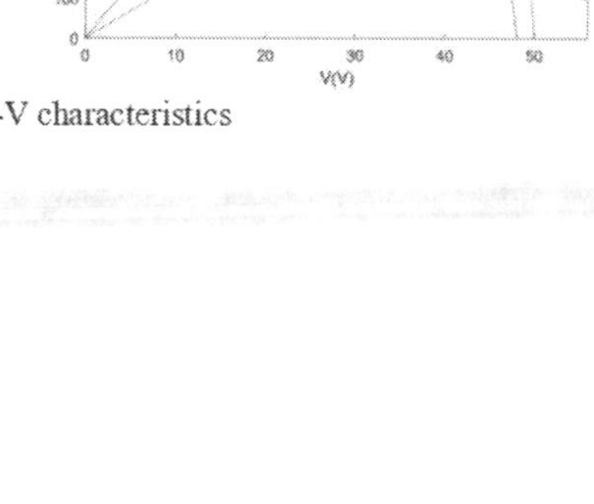

(a) P-V characteristics

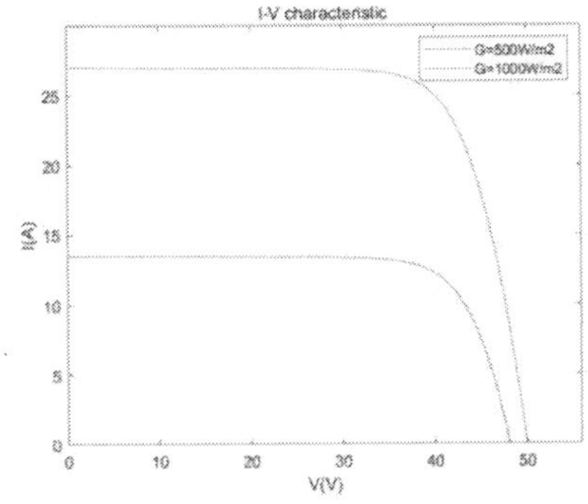

(b) V-I characteristics

Fig. 3. P-V and V-I characteristics when G=1000W/m2 and 500 W/m2.

From the P-V characteristic curves showing in Fig. 3, when irradiation is 1000W/m^2, the output power of the PV cells at the MPP (maximum power point) is around double the output power at MPP when G = 500W/m^2 (1000.6W output power at the MPP of a PV module when G=1000W/m^2 and the maximum power of 489.45W when G=500W/ m^2). And the voltages of MPP in both conditions are same (both 39.98V at MPP). For the I-V curve shown in Fig. 3, both MPPs with different irradiations that share the same voltage, which is also 39.98V, where the currents of both conditions are 25.02A and 12.24A respectively.

3 The Dual-loop Control for Providing Reactive power

Normally, the PV system running under unit power factor, but it is worth mentioning that control strategy can support reactive power when a low voltage fault occurs in AC bus without additional compensation device. The structure of PV system with three-phase voltage source inverter is presented in 3.1, which connected with an ideal three-phase voltage source. In 3.2, the typical dual-loop control for three-phase inverter I introduced. The proposed Q/V control strategy is expounded in 3.3. Finally, simulation results are discussed.

3.1 Three-phase Inverter Structure.

Transformation processes of PV arrays are mainly depending on the operations of grid-connected power converters and related components. The grid-connected inverter is the key component to convert DC voltage into the required AC voltage for supporting utility grid or loads. As shown in Fig. 4, a single-stage topology for PV system with a typical structure of three-phase grid-connected power inverter system is displayed where the filter capacitor placed in parallel with inverter and PV array.

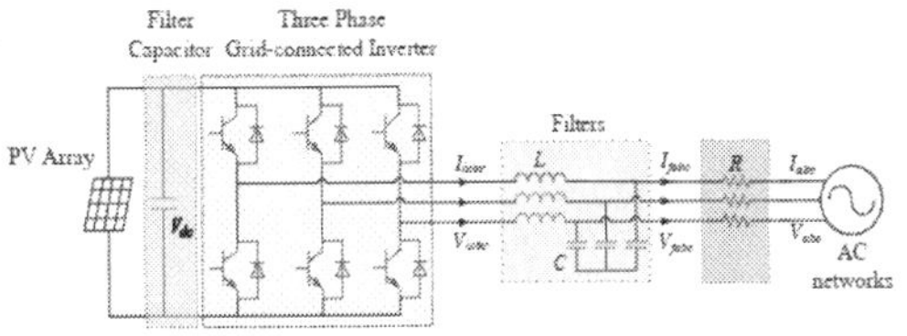

Fig. 4. Schematic diagram of the single-stage three-phase grid-connected inverter.

In this system, three-phase inverter in charge of all requirements for the appropriate three-phase AC voltage for utility grid and implementation of MPPT technology. Where V_{dc} stands for the DC side voltage; V_{iabc} and I_{iabc} stand for the *a*-, *b*- and *c*-phase output voltage and current respectively of the grid-connected inverter; V_{fabc} and I_{fabc} stand for the *a*-, *b*- and *c*-phase voltage and current that after the filters respectively. $L=L_a=L_b=L_c$ that stands for the single-phase inductor value of *LC* filter; *C* stands for the capacitor value for very single phase when that connected in the 'star' topology. $R=R_a=R_b=R_c$ that stands for the line resistance value, neglecting the line inductance.

Additionally, transformations and line models are neglected when analysis. Meanwhile, suppose three-phase voltage (V_{abc}) of the power grid, which represented by an AC source in the system, are balanced. The fundamental inverter equations about the voltage of the inductors and the capacitors' current are given in the following:

$$L\frac{dI_{iabc}}{dt}=V_{iabc}-V_{fabc} \\ =V_{iabc}-\left(V_{abc}+RI_{abc}\right) \quad (6)$$

$$C\frac{dV_{fabc}}{dt}=I_{iabc}-I_{abc} \quad (7)$$

For the active and reactive power decoupling control, the coordination transformations from three phases stationary abc reference to two-phase synchronous rotating dq reference frame are needed. By the Park transformations, the inverter equations of state can be written as:

$$\begin{cases} u_{id}=L\dfrac{di_d}{dt}+Ri_d-\omega Li_q+u_d \\ u_{iq}=L\dfrac{di_q}{dt}+Ri_d+\omega Li_q+u_d \end{cases} \quad (8)$$

Where u_d and u_q are respectively defined as the d-axis and q-axis component grid voltage, u_{id} and u_{iq} are represented as the d-axis and q-axis component out voltage of the three-phase inverter. id and i_q are defined as the d-axis and q-axis component of output current after filters that approximately equal to inverter current if capacitors can be neglected.

3.2 The Typical Dual-loop Decoupling Control for Three-phase PV Grid-connected Inverter

Fig. 5 gives a typical dual-loop control topology of three-phase grid-connected inverter. The entire control processes can be divided into: 1) parameters measurements from grid and transformation of coordinates processes; 2) phase locked loop (PLL) control; 3) outer-loop control or primary control; 4) inner-loop control and 5) phase width modulation (PWM) technologies.

Overall, in order to acquire certain output power by the control methods, firstly the system parameters should be measured for calculations such as active (P_{grid}) and reactive (Q_{grid}) power. And then measured powers and reference values are sent to outer-loop controller to generate reference current signals, which as reference signals for inner-loop control. The inner-loop controller aims to provide SPWW modulator with control signals, which reduces the burden on system regulator, increases system response speed and system dynamic performance.

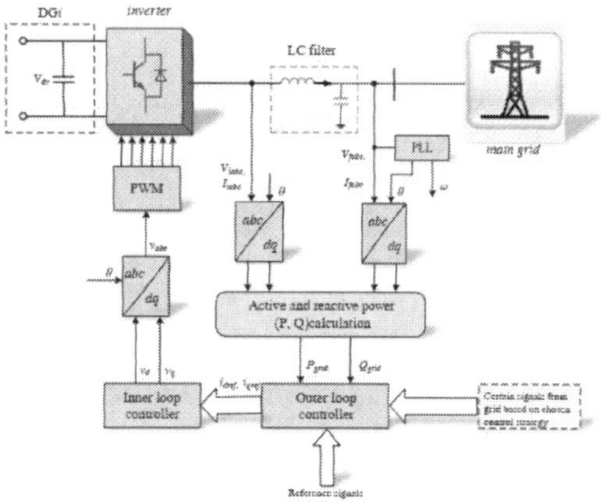

Fig. 5. A Typical Control Topology for Grid-connected Inverter.

Due to the Park transformation when a-phase is aligned with q-axis, the q-axis component of voltage equals to zero, and then power equations can be simplified, which is shown as follows:

$$\begin{cases} P_{grid} = \dfrac{3}{2}u_d i_d \\ Q_{grid} = -\dfrac{3}{2}u_d i_q \end{cases} \tag{9}$$

In this condition, active power only has relation with u_d, and reactive power only controlled by u_q. This is one type of the constant power control that under simplified mode.

3.2.1 Out-loop Controller.

For variable control purpose, different control strategies are employed in the outer-loop controller. Generally, in the PV system, trying to provide maximum output power, and working at unit power factor. Control process for active power is given in Fig.6. And i_{qref} is set to 0 directly for the unit power factor.

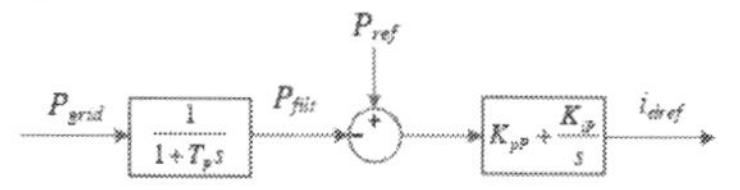

Fig. 6. Typical Outer-loop Control.

3.2.2 Inner-loop Controller.

From the above dq components in (8), it is difficult for designing controller due to the coupling of the d-axis and q-axis. The strategy of feed-forward decoupling control can be employed and introduce PI regulator to realize stable operation in the close-loop. The control relations are given in follows [6]:

$$\begin{cases} u_{id} = \left(K_P + \dfrac{K_I}{s}\right)\left(i_{dref} - i_d\right) - \omega L i_q + u_d \\ u_{iq} = \left(K_P + \dfrac{K_I}{s}\right)\left(i_{qref} - i_q\right) + \omega L i_q + u_d \end{cases} \tag{10}$$

The feed-forward decoupling control and coupling compensation can realize independent control of dq components. The diagram of the inner-loop current controller is shown in Fig.7.

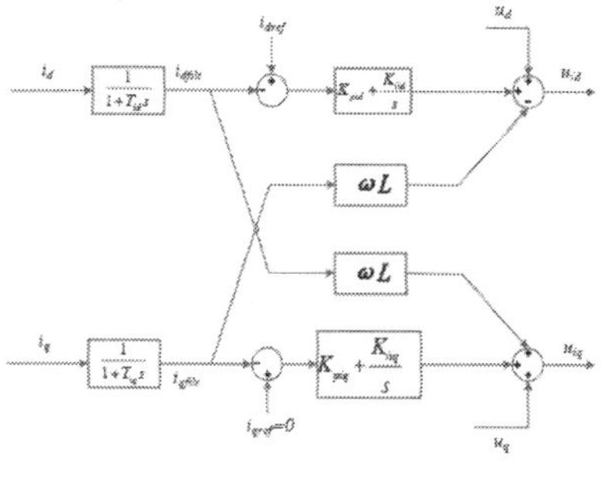

Fig. 7. Typical Inner-loop Control.

3.3 Reactive power/Voltage Control in PV System.

Sometimes PV arrays are required to provided certain reactive power for voltage fluctuation or faults, and improve low-voltage ride through capability (LVRT) [7]. A reactive control topology is employed to provide certain output reactive power. Due to the P-V parameters that displayed in Fig.3, a strategy by means of controlling output PV array voltage (V_{dc}, shown in Fig. 5) to regulate output power (P) is introduced. The maximum power point tracking (MPPT) algorithm is used to find operation voltage V_{mpp} for PV arrays that corresponds with the maximum output power. And then control V_{mpp} as the reference signal, which can be shown in Fig.8, to regulate V_{dc} as well as P [8]. The generated signal i_{dref} is also sent to inner-loop controller to regulate P. Reactive power (Q) can also be controlled by PI controller where the i_{qref} is sent to inner-loop controller. While the inner control is as same as the typical dual-loop control method.

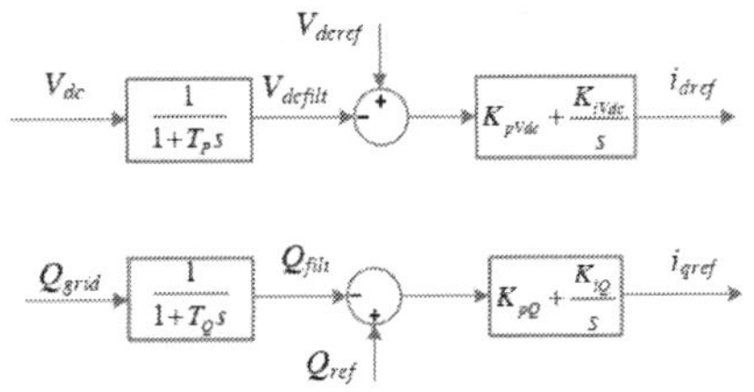

Fig. 8. Typical Q/V Control.

4 Simulation Results

For the simulation model of the system, the resistance for per phase in the balanced lines is $0.0001\,\Omega$, inductance value is $125\,\mu H$, The rms (root mean square) value of phase to phase voltage for the main grid is 230 V.

In case 1, for the nominal condition of PV arrays, setting $i_{qref} = 0$, which means running at unit power factor. While in case2, based on the decoupling control and Q/V control strategy, active power is still required, additional reactive power is expected. According to (9), setting Q_{ref}=28.8 kVar. The active power of PV arrays still output the maximum values under its external conditions. Besides, in order to test the dynamic working conditions of the PV system, in the both cases, solar irradiation changed from 1000 W/m^2 to 500 W/m^2 at 1 second.

According to the simulation structure, the output active power and reactive power in grid side under both control topology are measured for testing, which are given in Fig. 9.

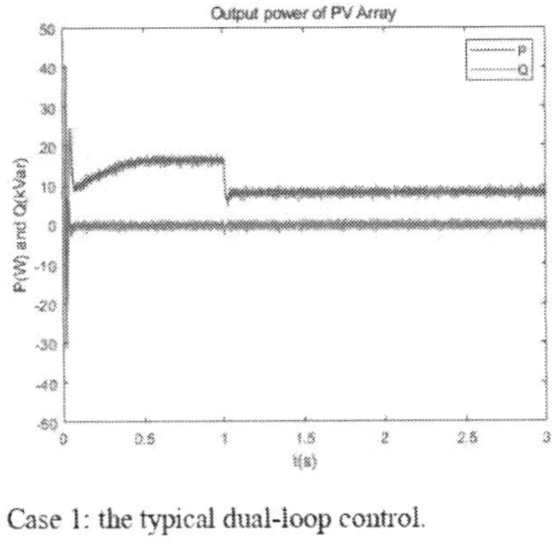

(a) Case 1: the typical dual-loop control.

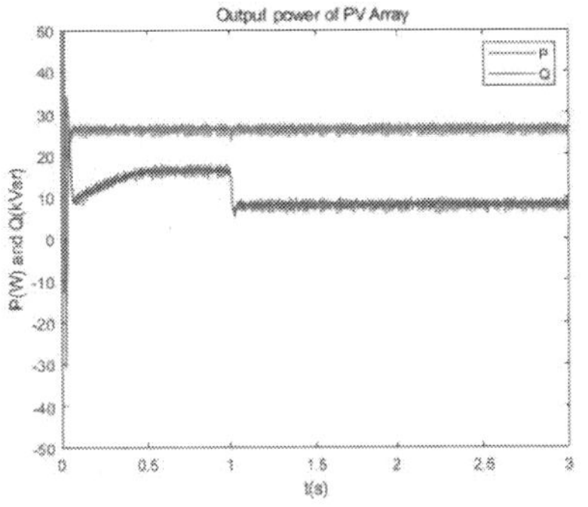

(b) Case 2: the Q/V control strategy.

Fig. 9. Curves of P and Q in different control conditions.

- In both cases, at 1 second, the output power deduces half from 17.97 kW to 8.95 kW when irradiation changed from 1000 W/m^2 to 500 W/m^2, which shows quite perfect dynamic performance of active power control of PV arrays.
- In case1, when setting $i_{qref} = 0$, it can directly control output zero reactive power, which realizes working in unity power factor.
- Compared with simulation results in case1 and case2, since the reactive power control, blue lines (reactive power) changed, but red lines (active power) still maintained maximum output power under the control order. Thus, it verified the effectiveness and accuracy of decoupling control.
- In case2, it is shown that the reactive power supporting in Fig. 9(b), the $i_{qref} = -1.2$, according to (9), and the base value of Q is 16 kVar, as a result, the calculated value of Q in grid side is given.

$$Q_{grid} = -\frac{3}{2}u_d i_q = -\frac{3}{2}\times 1\times(-1.2)\times 16 = 28.8\ kVar$$

- The curves of active power (P, red line) in case1 and case2 are almost same, which verified the strategy is feasible that regulate P by means of controlling V.

5 Conclusions

The simulation results show that the output active power is quite perfect even in case of irradiation changes. On the other hands, reactive power is required in many conditions and for the enhancements of low-voltage ride through capability. The proposed reactive power control method that based on the decoupling dual-loop control strategy can provide certain reactive power effectively. Simulation results verify the effectiveness and accuracy of the Q/V control.

Project Supported by National Natural Science Foundation of China (NSFC) (51761145106).

References

1. Z. Gao, S. Li, X. Zhou, and Y. Ma. An overview of PV system, *IEEE International Conference on Mechatronics and Automation*, 587–592 (2016)
2. J. Rocabert, A. Luna, F. Blaabjerg, and P. Rodríguez. Control of Power Converters in AC Microgrids. IEEE Trans. Power Electron, **27**, 4734–4749 (2012)
3. M. G. Villalva, J. R. Gazoli, and E. R. Filho. Comprehensive Approach to Modeling and Simulation of Photovoltaic Arrays. IEEE Trans. Power Electron, 24, 1198–1208 (2009)
4. Y. Jiang, J. A. A. Qahouq, and I. Batarseh. Improved solar PV cell Matlab simulation model and comparison. In: Proc. of 2010 *IEEE International Symposium on Circuits and Systems*, 2010, pp. 2770–2773.
5. H.-L. Tsai, C.-S. Tu, and Y.-J. Su, Development of Generalized Photovoltaic Model Using MATLAB/SIMULINK', **6** (2008)
6. T. Huang, X. Shi, Y. Sun, and D. Wang. Three-phase photovoltaic grid-connected inverter based on feedforward decoupling control. *2013 International Conference on Materials for Renewable Energy and Environment*, **2**, 476–480 (2013)
7. A. V. Timbus, R. Teodorescu, F. Blaabjerg, M. Liserre, and P. Rodriguez. Linear and Nonlinear Control of Distributed Power Generation Systems. *2006 IEEE Industry Applications Conference Forty-First IAS Annual Meeting*, **2**, 1015–1023 (2006).
8. T. Huang, X. Shi, Y. Sun, and D. Wang. Three-phase photovoltaic grid-connected inverter based on feedforward decoupling control. *International Conference on Materials for Renewable Energy and Environment*, **2**, 476–480. (2013)

https://doi.org/10.1051/e3sconf/202018203001

Energy Management of Hybrid Power System in UUV Based on Pontryagin's Minimum Principle

Huizhong Gao, *Junguang* Wang, *Jun* Lu *, *Xiao* Zong

Xi'an precision mechanical research institute, 710077, Xi'an PR China

Abstract. Hybrid power system which consists of a proton exchange membrane fuel cell and a battery has great perspective for Unmanned Underwater Vehicles (UUVs). As fuel cell's working point has a direct relationship with its efficiency and lifetime, a key concern of the hybrid power system is the power distribution between the two parts. Aiming at minimizing the equivalent hydrogen cost, this work presents an optimal real-time energy management strategy based on the Pontryagin's minimum principle. Compared with some other methods, this method possesses shorter calculation time and smaller storing space, which is helpful for online optimization in actual applications. In this research, a typical UUV working profile is tested with different initial co-state values to evaluate the proposed method, and the results indicate good robustness and effectiveness.

1 Introduction

In the past decades, many countries have put a lot of efforts to evolve underwater unmanned vehicles (UUVs) from lab research into commercial, military and scientific applications[1]. With the outstanding advances, UUVs have become the key role in ocean development. Generally, UUVs are characterized with extended range, long endurance, and low noise. Current fleet of UUVs is usually battery powered [2][3] as batteries are a simple solution. This factor combined with a battery's silence and ability to turn on and off instantly make it very attractive for UUVs. However, the potential for battery to enhance its energy density is quite limited. As such, battery technology cannot catch up with the UUV development with extended range and longer endurance.

A hybrid power system in UUVs often takes fuel cell as the main power source and battery as the auxiliary power source. In this condition, the energy management strategy (EMS) is required to split the power between the two sources [4]. For better performance, EMS should take the load demand and resource constraints into consideration. Ideally, by controlling the DC/DC converter's output current, EMS ensures the hybrid power system works at high efficiency. Previously, various EMS, such as rule-based strategies [5], instantaneous/global optimized strategies [6], intelligent algorithms [7][8] are studied. Pontryagin's minimum principle (PMP) is an optimal control technique firstly proposed by Pontryagin, which transforms the global optimization problem described by DP to an instantaneous Hamiltonian optimization problem. It doesn't need the whole cycle as prior information and can find the quasi optimal solution, which is proper to solve UUVs' online energy management problem.

In this paper, PMP-based EMS is studied for UUV application. Section 2 describes the structure of the fuel cell hybrid power system. Section 3 makes an introduction of the PMP theory, and raises the energy management problem. Then, in section 4, PMP strategy is used to achieve the solution, and simulation test is conducted to verify the method. Eventually, the conclusion is given in section 5.

2 The hybrid powertrain

There are several powertrain topologies as candidates for hybrid power system. As the fuel cell is efficient and the battery has good dynamics, in this paper, a unidirectional DC is equipped on the fuel cell side and the battery is connected on bus directly. The whole powertrain is illustrated in Fig.1.

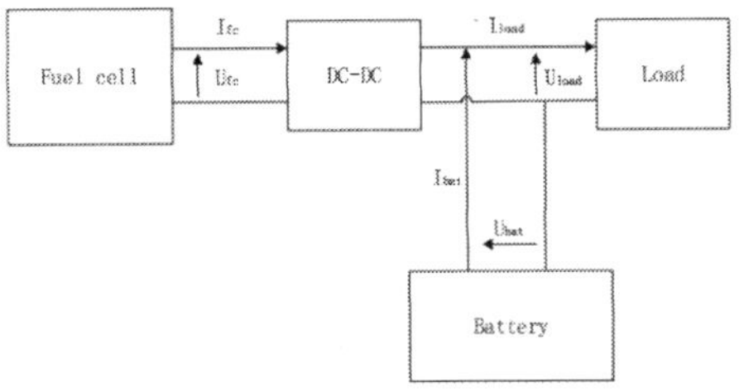

Fig.1. The topology of the FC hybrid power system

2.1 Fuel cell model

* Corresponding author: lujun_705@163.com

Different from the hydrogen-air FC in buses, the hydrogen-oxygen FC in UUVs take the pure oxygen as oxidant. Therefore, the partial pressure in cathode is relevantly different. In [9], the author makes a comprehensive study on FC model.

According the electric chemical theory of FC[10], a single cell output voltage can be defined as a combination of the thermodynamic potential E_{nernst} , the activation overvoltage V_{act} and the ohmic overvoltage V_{ohm} , coefficients of activation over voltage ξ_i .

$$V_{cell} = E_{energy} - V_{act} - V_{ohm} \tag{1}$$

where

$$E_{nernst} = \begin{pmatrix} E_0 - 0.85*10^{-3}(T_{body} - 298.15) \\ + \dfrac{RT_{body}}{2F} \ln[p_{H_2} * (p_{O_2})^{0.5}] \end{pmatrix} \tag{2}$$

$$V_{act} = -\begin{pmatrix} \xi_1 + \xi_2 T_{body} + \xi_3 T_{body} \ln C_{O_2} \\ + \xi_4 T_{body} \ln I \end{pmatrix} \tag{3}$$

$$\xi_2 = \begin{pmatrix} 0.00286 + 0.0002 \ln A_{cell} \\ + 4.3*10^{-5} \ln C_{H_2} \end{pmatrix} \tag{4}$$

$$C_{H_2} = 9.174*10^{-7} p_{H_2} \exp(\frac{-77}{T_{body}}) \tag{5}$$

$$C_{O_2} = 1.97*10^{-7} p_{O_2} \exp(\frac{498}{T_{body}}) \tag{6}$$

$$V_{ohm} = I * R_{int} \tag{7}$$

$$R_{int} = \frac{r_{mem} * t_{mem}}{A_{cell}} \tag{8}$$

$$r_{mem} = \frac{181.6[1 + 0.03J + 0.062(T_{body}/303)^2 J^{2.5}]}{[\gamma - 0.634 - 3J]\exp[4.18((T_{body} - 303)/T_{body})]} \tag{9}$$

Additionally, the saturated vapour pressure $p_{H_2O}^{sat}$ can be calculated by the operating temperature , and the effective partial pressure p_{H_2} and p_{H_2} are described as follows:

$$Log(p_{H_2O}^{sat}) = \begin{pmatrix} 2.95*10^{-2}(T_{body} - 273.15) - \\ 9.18*10^{-5}(T_{body} - 273.15)^2 + \\ 1.44*10^{-7}(T_{body} - 273.15)^3 - 2.18 \end{pmatrix} \tag{10}$$

$$p_{H_2} = 0.5 p_{H_2O}^{sat}[\frac{1}{X_{H_2O}^{channel} \exp(\frac{1.653J}{T_{body}^{1.334}})} - 1] \tag{11}$$

$$p_{O_2} = p_{H_2O}^{sat}[\frac{1}{X_{H_2O}^{channel} \exp(\frac{4.192J}{T_{body}^{1.334}})} - 1] \tag{12}$$

Based on the analysis all above, the FC stack output voltage is calculated by

$$V_{stack} = N_{cell} * V_{cell} \tag{13}$$

2.2 Battery model

The Rint model is widely used to describe the performance of the battery as showed in Fig.2, in which the battery is transferred into a voltage source and a resistor in series.

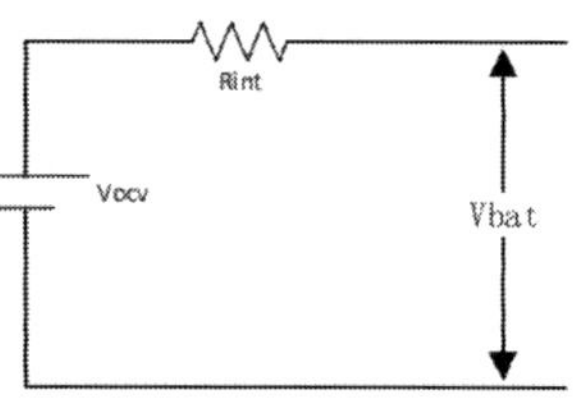

Fig.2. The Rint model

Usually the variation of the internal resistance of charge and discharge of battery has been related to the state of charge (SOC). Generally the current and the voltage can be expressed as below, when the battery power is P,

$$I_{bat} = \frac{V_{ocv} - \sqrt{V_{ocv}^2 - 4R_{int}P}}{2R_{int}} \tag{14}$$

$$V_{bat} = V_{ocv} - R_{int} I_{bat} \tag{15}$$

where

$$V_{ocv} = \begin{cases} f_1(SOC), charge \\ f_2(SOC), discharge \end{cases} \tag{16}$$

$$R_{int} = \begin{cases} f_3(SOC), charge \\ f_4(SOC), discharge \end{cases} \tag{17}$$

Furthermore, the battery SOC can be achieved by

$$SOC = SOC_0 - \int_0^t \frac{I_{bat}}{Q_{bat}} dt \tag{18}$$

where Q_{bat} is the capacity of the battery.

3 Energy management of hybrid power system based on PMP algorithm

3.1 Pontryagin's minimum principle

The Pontryagin's minimum principle (PMP) is a very valid approach in optimal control theory. Although not sufficient, every solution that satisfies the necessary conditions is called an extremal solution. Obviously, if the optimal solution exists, it must be one of the extremal solutions.

Generally, the optimal control problem is defined to seek an accessible control variable u^* that lead the system $x(t) = f[x(t), u(t), t]$ to follow a certain trajectory x^* that minimizes the performance index J.

$$J(u) = \begin{pmatrix} \Phi[x(t_f), t_f] + \\ \int_{t_0}^{t_f} L[x(t), u(t), t]dt \end{pmatrix} \tag{19}$$

On one hand, $L[x(t), u(t), t]$ is the instantaneous cost function and its integration is integral cost, which decide the performance of $x(t)$ and $u(t)$. On the other hand, $\Phi[x(t_f), t_f]$ is the terminal cost which can evaluate the terminal state of the system. In PMP theory, furthermore, a Hamilton function is given,

$$H[x(t), u(t), \lambda(t), t] = \begin{pmatrix} L[x(t), u(t), t] + \\ \lambda^T(t) f[x(t), u(t), t] \end{pmatrix} \tag{20}$$

where $\lambda(t)$ is a Lagrange multiplier called co-state. Then, the following relations are established:

$$\dot{x}^*(t) = \frac{\partial H}{\partial \lambda(t)}[x^*(t), u^*(t), \lambda(t), t] \tag{21}$$

$$\dot{\lambda}(t) = -\frac{\partial H}{\partial x^*(t)}[x^*(t), u^*(t), \lambda(t), t] \tag{22}$$

$$H[x^*(t), u^*(t), \lambda(t), t] \le H[x(t), u(t), \lambda(t), t] \tag{23}$$

3.2 Energy management based on PMP

UUV works in the underwater where the fuel/oxidizer cannot be acquired from the outside environment. It is critical for EMS to take optimal utilization of the on-board energy reserves. Because the battery charging-discharging process may cause energy loss, the battery should be charged by fuel cell as less as possible to reduce the energy loss. Therefore, the objective function of the UUV's EMS is defined as

$$J = \min \sum (\dot{m}_{bat} + \dot{m}_{fc}) \tag{24}$$

Considering the system features and the analysis above, the state equation is described as follow,

$$\dot{x} = \dot{SOC} = -\frac{I_{bat}}{Q_{bat}} \tag{25}$$

$$I_{bat} = \frac{V_{oc} - \sqrt{V_{oc}^2 - 4R_{int} P_{bat}}}{2R_{int}} \tag{26}$$

where Q_{bat} is the battery capacity, I_{bat} is the battery current and V_{oc} is the open circuit voltage of the battery.

The equivalent hydrogen consumption of battery $\dot{m}_{bat}$ is a function of battery power P_{bat}

$$\dot{m}_{bat} = P_{bat}(\eta_{chg}\eta_{dis})^{-\mathrm{sgn}(P_{bat})} \tag{27}$$

To solve the allocation problem of hybrid power system, according to the PMP, the Hamilton function is defined below,

$$H = \dot{m}_{bat} + \dot{m}_{fc} - \lambda * \frac{I_{bat}}{Q_{bat}} \tag{28}$$

$$\dot{SOC}(t) = \frac{\partial H}{\partial \lambda} = -\frac{V_{oc}(SOC) - \sqrt{V_{oc}^2(SOC) - 4R_{int}P_{bat}}}{2R_{int}Q_{bat}} \tag{29}$$

$$\dot{\lambda}(t) = -\frac{\partial H}{\partial SOC} = \frac{\lambda(t)}{Q_{bat}} \begin{pmatrix} \frac{\partial I_{bat}}{\partial V_{oc}} * \frac{\partial V_{oc}}{\partial SOC} \\ + \frac{\partial I_{bat}}{\partial R_{int}} * \frac{\partial R_{int}}{\partial SOC} \end{pmatrix} \tag{30}$$

Additionally,

$$\dot{\lambda}(t) = -\frac{\lambda(t)}{Q_{bat}} * \frac{I_{bat}}{\sqrt{V_{oc}^2(SOC) - 4R_{int}P_{bat}}} * \frac{\partial V_{oc}}{\partial SOC} \tag{31}$$

At any instant, there will be an optimal control variable P_{dc}^* makes H reaches its minimum value

$$P_{dc}^* = \arg\min H \tag{32}$$

Based on this, the corresponding battery power P_{bat}^* can be easily calculated with corresponding constraints

$$P_{bat}^* = P_{req} - P_{dc}^*, P_{bat} \in [P_{bat,\max}, P_{bat,\min}] \tag{33}$$

3.3 Energy management problem solution

Commonly, the PMP-based optimal problem can be converted into a two-point boundary value problem, and utilize a numerical method to solve the optimization of the Hamiltonian function. Shooting method is a widely used numerical approach in differential equation solving, whose cardinal process is to seek out optimal initial co-state λ_0^* iteratively. In other words, only the proper selected λ_0^* can lead to a satisfied shooting process. By using the Secant method, the initial co-state is modified at each shooting process until the condition SOC(t_f)=SOC_L is fulfilled, whose form is

https://doi.org/10.1051/e3sconf/202018203001

$$\begin{cases} \lambda_1 = \lambda_0, i = 1 \\ \lambda_2 = \lambda_1 + \delta, i = 2 \\ \lambda_i = \lambda_{i-1} - (\lambda_{i-1} - \lambda_{i-2}) * \dfrac{SOC_{i-1,t_f} - SOC_{t_f}}{SOC_{i-1,t_f} - SOC_{i-2,t_f}}, i = 3,4,\dots \end{cases} \tag{34}$$

where λ_i is the initial co-state value in i-th shooting process, SOC_{i,t_f} is the end SOC in i-th shooting process.

During the solving procedure, the SOC and λ are changed along with their derivatives in the form of

$$SOC_i^{k+1} = SOC_i^k + \dot{SOC} * \Delta t \tag{35}$$

$$\lambda_i^{k+1} = \lambda_i^k + \dot{\lambda} * \Delta t \tag{36}$$

Eventually, an intelligent iterative strategy based on instantaneous data by trial and error over a several simulations and contrast is able to be implemented.

4 Simulation and discussion

The hybrid UUV usually works in periodic power cycle and supplies enough energy and power in its mission. In order to verify the effect of PMP-based energy management strategy, a typical power profile of small-sized UUV is simulated in this section, as depicted in Fig.3. The specifications of the fuel cell/battery hybrid power system and the PMP parameters are showed in table 1.

Table 1. System parameters

Parameter item	value
Fuel cell max power	5000W
Battery max power	3000W
Capacity	9.6Ah
SOC upper level	90%
SOC lower level	25%
ΔP	1W
Δt	1s
λ_0	-10/-15/-20
δ	-4
E	0.001

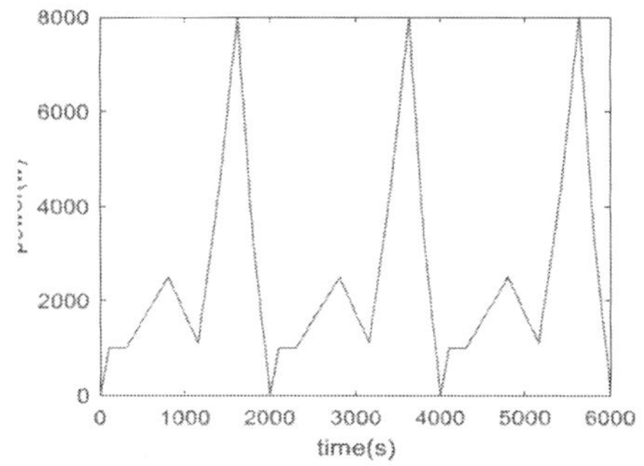

Fig.3. Power profile

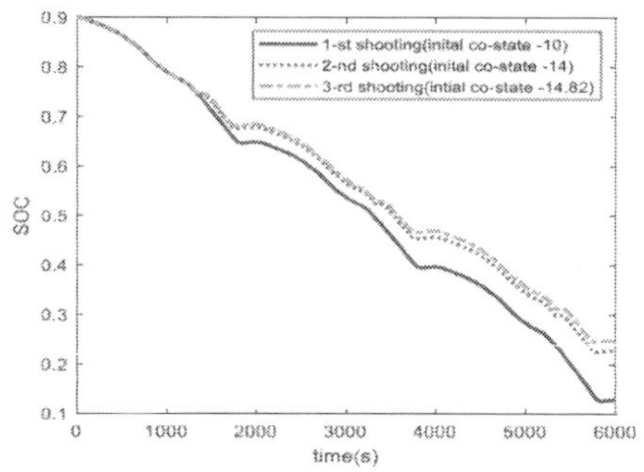

Fig.4. SOC trajectory of $\lambda_0 = -10$

Fig.3 shows the power profile, meanwhile, the optimal trajectories of SOC are illustrated in Fig. 4-6.

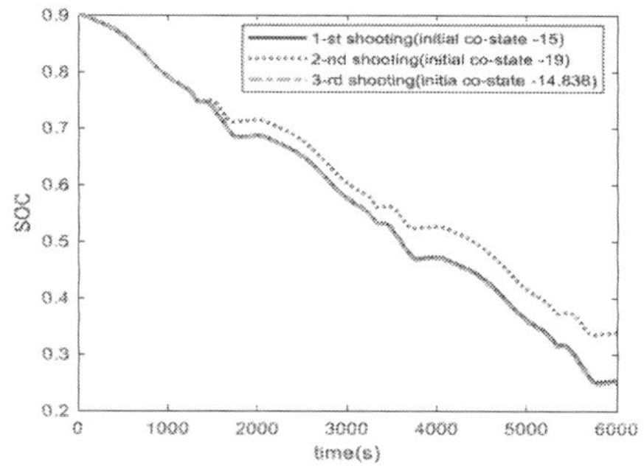

Fig.5. SOC trajectories of $\lambda_0 = -15$

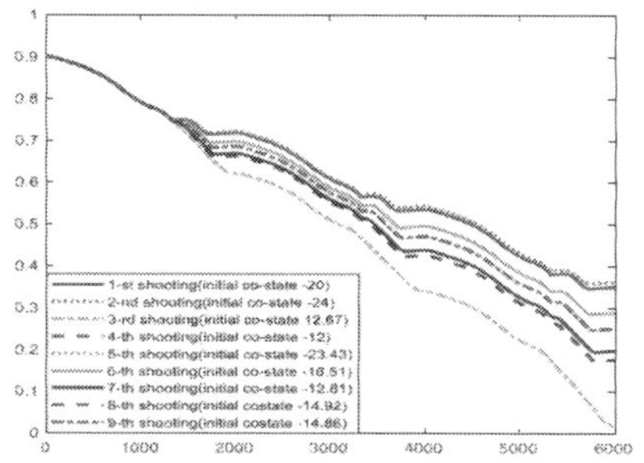

Fig.6. SOC trajectories of $\lambda_0 = -20$

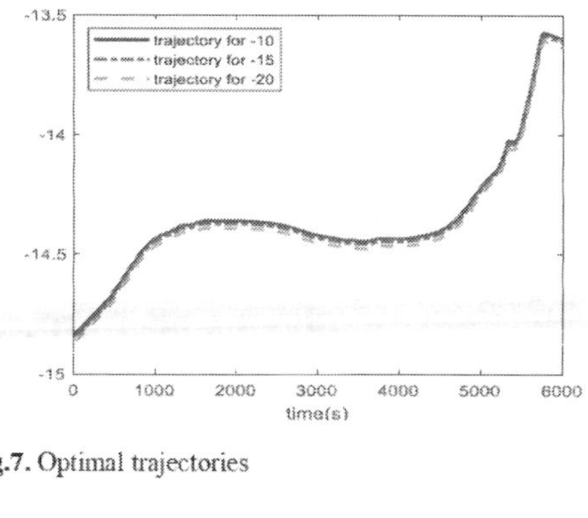

Fig.7. Optimal trajectories

In the presented algorithm, by precisely regulating the power distribution during the mission process, the battery SOC is well decline from the initial value to a permitted minimum. The results in Fig.4-Fig.6 indicate that, although the shooting sequences start from different initial co-state value, they can eventually reach almost the same optimal co-state value (-14.825 for $\lambda_0 = -10$,-14.837 for $\lambda_0 = -15$, and -14.858 for $\lambda_0 = -20$), specifically, the optimal co-state in the case of 3 distinct initial value that lead SOC change from 90% to 25% is also shown in Fig.7. Consequently, the optimal energy split path can be achieved through the optimal co-state and the trajectory.

5 Conclusion

This paper studies PMP and applies it to EMS of hybrid power system in UUVs. In the simulation, UUV's power profile is utilized to verify the optimal energy management strategy and a set of conclusions can be summarized as follow:

- The numerical solution of the PMP approach is achieved by using shooting method, which converts a constrained optimization problem into a two-point boundary value problem. In the calculation process, the Secant method can rapid tune the co-state λ to the optimal values, which satisfy both the power and SOC demand and minimize the consumption.
- The presented algorithm is examined by a simulated power profile with 3 different initial co-state values. The results indicate that the identical problem with different initial values has totally distinct calculation process, but can eventually reach the similar optimal solution. Therefore, the PMP-based EMS is robust and can offer an effective online solution for the UUVs' hybrid power system.

References

1. Vukic.*http://www.aig.si/13/clanki/vabljena_predavanja/U3_Vukic.pdf.* (2013).
2. Alam, K., Ray, T., Anavatti, S.G. *Ocean.* **88**(2014), 627–630.
3. Alaaeldeen M.E.Ahmed, Duan Wen-yang. *Journal of ship mechanics*, 2016, **20**,6: 768-787(2016).
4. A. Tahri, H. El Fadil, F.Z. Belhaj, et al. Manegement of fuel cell power and supercapacitor state-of-charge for electric vehicles.Electric Power Systems Research, 160:89-98(2018).
5. Schouten NJ, Salman MA, Kheir NA. *IEEE T Contr Syst T* ,**10**,3:460-468(2002).
6. Xu L, Li J, Hua J, Li X, Ouyang M. *Int J Hydrogen Energy* , **34**:7323-33(2009).
7. Xiaohui Zhang, Li Liu , Yueling Dai, et al. *International journal of hydrogen energy*, **43**:10094-10103(2018).
8. Hou C , Ouyang M , Xu L , et al. *Applied Energy*, **115**:174-189(2014).
9. Chien-Hsing Lee, Jian-Ting Yang. *Journal of Power Sources*, **196**: 3810-3823(2011).
10. Wang C , Nehrir M H , Shaw S R . *IEEE Transactions on Energy Conversion*, **20**,2:442-451(2005).

Power Reconstructing Method of Distributed Photovoltaic Based on the Temporal and Spatial Correlation

Shangqiang Li[1,*], *Rongfu* Sun[2], *Ying* Qiao[1], and *Zongxiang* Lu[1]

[1]State Key Lab of Control and Simulation of Power Systems and Generation Equipment, Department of Electrical Engineering, Tsinghua University, 100084 Haidian District in Beijing, China
[2]State Grid Jibei Electric Power Company, 100031 Xicheng District in Beijing, China

Abstract. 55GW distributed photovoltaic have been installed in China, but nearly half are connected to the low voltage level of 380V, without real-time power data acquisition. The sequential power data is needed to be reconstructed based on some related monitoring data. Current researches focus on outliers recovery, but not reconstruction from none. This paper explores the temporal and spatial correlation of power between adjacent centralized photovoltaic stations and proposes a large-scale missing power data reconstructing method based on the time-delay power correlation, the spatial geometric characteristics of stations and the thought of ensemble learning. Finally, we verify the effectiveness of the proposed method by simulation based on the real photovoltaic power data. The proposed method can get the better effect of data reconstructing compared with the traditional method, which only use the power curves of the nearest CP station to reconstruct the power curves of the DP station according the capacity conversion.

1 Introduction

Until the end of 2018, 55GW distributed photovoltaic (DP) have been connected with distribution network in China, which challenge the operation of distribution network a lot [1-4]. To make clear characteristics from historical data is the very start of solving problems.

One of the biggest challenges in China is the worse data conditions [5-6]. There are half of DP are connected to the voltage level of 380V, which have no real-time power data acquisition. For example, 30% of DP installation in northern Hebei Province of China can only collect every electricity energy data, but not power data. In addition, the available meteorological data of DP is very few and low precision due to the high economic cost. Therefore, we need to reconstruct the large-scale missing data of DP.

Actually, some studies have been carried out in outliers recovery. Literatures [7-10] reconstructed the missing power data based on principal component analysis and neural network, using the data of CP stations around the target DP station and related weather factors. The current studies on the data reconstructing depend on the meteorological data and more suitable for repairing the small-scale missing data. In addition, some intelligent algorithms like neural network require data labels, which means that they cannot handle data reconstructing from none. However, we can learn from their analysis methods of the temporal and spatial characteristics between different photovoltaic stations.

As for the large-scale missing power data reconstructing, the traditional method in engineering is currently only using the power curves of the nearest CP station to reconstruct the power curves of the DP station according the capacity conversion. It is obvious that this method would bring great errors.

Therefore, this paper, based on the real power data of centralized/distributed photovoltaic power stations, proposes daily electric energy ratio, power cross-correlation coefficient, time-delay power cross-correlation coefficient and other indicators to mine the temporal and spatial characteristics between CP and DP from the perspective of historical power data. This paper also proposes a data reconstructing method based on the time-delay power correlation, the spatial geometric characteristics of stations and the thought of ensemble learning, which can get smaller errors than the traditional data reconstructing method.

2 Characteristic mining and data reconstruction

Without historical power curves in DP, it is impossible to reconstruct the missing data using relatively advanced learning algorithms of supervised learning. For the data reconstructing of DP, we can only use its daily electrical energy and the historical power curves of peripheral CP. Therefore, we adopt the idea of unsupervised learning, based on the time-delay power correlation, the spatial geometric characteristics of DP and CP, to reconstruct the missing data.

2.1 Definition of characteristic indicators

* Corresponding author: 363524255@qq.com

In order to explore the temporal and spatial characteristics of CP and DP historical power curves, we proposed station relative distance, daily electrical energy, ratio of daily electrical energy, power cross-correlation coefficient and time-delay power cross-correlation coefficient in this section.

Station relative distance (km) refers to the relative distance of each photovoltaic power station, which is calculated by latitude and longitude coordinates, reflecting the spatial characteristics of each station, as shown in (1). In the formula, the lg_1 and lg_2 represent longitude of two stations respectively. The la_1 and la_2 represent latitude of two stations respectively.

(1)

Daily electrical energy refers to the sum of the daily power curve of a photovoltaic station, as shown in (2). In the formula, the n represents the name of a photovoltaic station. The d represents one day. The M represents total number of data points in a power curve. The $P_{n,d}(i)$ represents the active power of the station n at time of i on the day d.

$$Q_n(d)=\sum_{i=1}^{M}P_{n,d}(i) \tag{2}$$

Ratio of daily electrical energy refers to the ratio of days in which daily electric energy of CP is larger than that of DP, as shown in (3). In the formula, the D represents the total number of statistical days. The Q_{DP} represents daily electric energy of DP.

$$ratio_{n,DG}=\frac{\sum_{d=1}^{D}f(Q_n(d))}{D}$$
$$f(Q_n(d))=\begin{cases}1, Q_n(d)\ge Q_{DP}(d)\\ 0, others\end{cases} \tag{3}$$

Power cross-correlation coefficient refers to the correlation coefficient between a daily power curve of CP and that of DP, as shown in (4). In the formula, $P_n(d)$ and $P_{DG}(d)$ represent the power sequence of CP station n and DP station on the day d respectively. It should be noted that we only use the power sequence in 4 hours around 12 noon to calculate the correlation coefficient, which can reflect the characteristic of a photovoltaic power curve better.

$$\rho_{n,DG}(d)=\frac{\mathrm{cov}(P_n(d),P_{DG}(d))}{\sigma_{P_n(d)}\sigma_{P_{DG}(d)}} \tag{4}$$

Time-delay power cross-correlation coefficient refers to the correlation coefficient between a power curve of CP after a certain distance of time translation and that of DP, as shown in (5). In the formula, the dt represents the time translation distance. The $x_{\max}$ represents the maximum of the time translation distance, which should ensure that the power sequence used for calculation after time translation belongs to the same day.

$$\rho_{\mathrm{delay},n,DG}(d)=\frac{\mathrm{cov}(P_n^{'}(d,\mathrm{dt}),P_{DG}(d))}{\sigma_{P_n^{'}(d,\mathrm{dt})}\sigma_{P_n(d)}}, d=1,2,\ldots,D$$
$$P_n^{'}(d,\mathrm{dt})=P_n(d,t+dt),$$
$$dt\in\{x\mid -x_{\max}\le x\le x_{\max}, x\in Z, x_{\max}\ge 0\} \tag{5}$$

2.2 Characteristics mining

In order to explore the temporal and spatial characteristics of CP and DP historical power curves, we select a DP station and 8 adjacent CP stations and take their historical power as the object of analysis. Fig. 1 shows their relative position and capacity data.

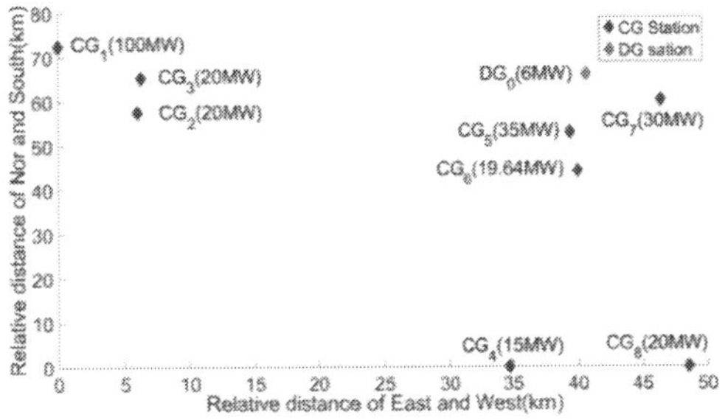

Fig. 1. Spatial Distribution Chart of Photovoltaic Stations.

(1) Characteristics about Electrical Energy

In order to compare the differences about *daily electrical energy* between DP and CP, we firstly convert the daily power curves of CP stations according to the capacity of DG_0 by (6). In the formula, S_n represents the capacity of CP station n.

$$P_{nc}(d)=P_n(d)/S_n\times S_{DG_0}=P_n(d)/S_n\times 6 \tag{6}$$

Furthermore, we calculate the *ratio of daily electrical energy* by (3) to get Fig. 2. We can know that the *ratio of daily electrical energy* of CP stations is close to 1 except for CG_6, which is caused by the better photovoltaic maintenance of CP stations than GP stations. However, the reason that the *ratio of daily electrical energy* of CG_6 is close to 0 may be because some inverters of CG_6 are not in operation.

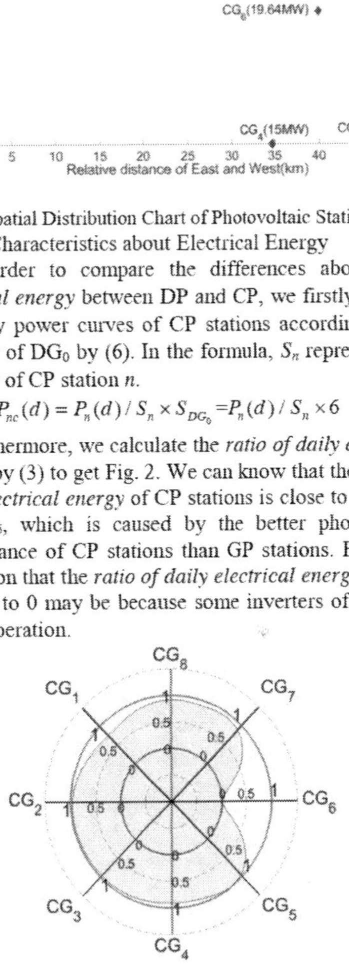

Fig. 2. Analysis Chart of Ratio of Daily Electrical Energy.

In general, under the same installed capacity, there are great differences between the *daily electrical energy* of CP and DP. Thus, when we use the power data of CP to reconstruct the missing power data of DP, we should consider the differences about electrical energy rather than the direct capacity conversion.

(2) Characteristics about Correlation

Time-delay power cross-correlation coefficient reflects temporal characteristics of power between different stations. Fig. 3 shows the *time-delay power cross-correlation coefficients* for four typical days of CG_7 and the corresponding power curves of CG_7 and

DG_0. The power curves of CG_7 are converted according to the capacity of DG_0. We can know that the correlation coefficient increases first and then decreases with the increase of time delay, and reaches its maximum on one point called the best delay time. However, this trend would weaken with the increase of power curve fluctuation. When the power curves is smooth, maybe in a sunny day, the station relative determines the best delay time. When the power curve is smooth, maybe in a sunny day, the best delay time is determined by the station relative. When the power curve fluctuates greatly, maybe in an overcast day, the best delay time may also be influenced by the speed of wind or other weather conditions.

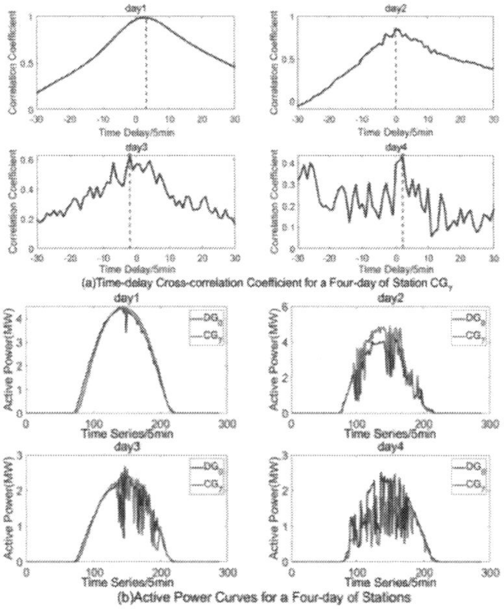

Fig. 3. Analysis Chart of Time-delay Power Cross-correlation Coefficient.

Furthermore, we calculate the average *time-delay power correlation coefficients* of each CP station, as shown in Fig. 4. In general, the time-delay correlation between CP power curves and DP power curves is obvious, which can be used for data reconstructing.

(3)Other Characteristics

When we do data reconstructing, we don't know the power data of DP stations, so that we can't directly use its power curves to calculate the *time-delay power cross-correlation coefficient*. In order to use time-delay characteristic to reconstruct data, we need to find other power curves similar to the missing power curves. Thus, we calculate the correlation coefficients between the daily power curves of other CP stations and those of CG_7. Compared this correlation coefficient and *power cross-correlation coefficient* calculated before, we can know that for the different CP stations, The ratio of days where their difference is less than 0.1 is shown in the Table 1.

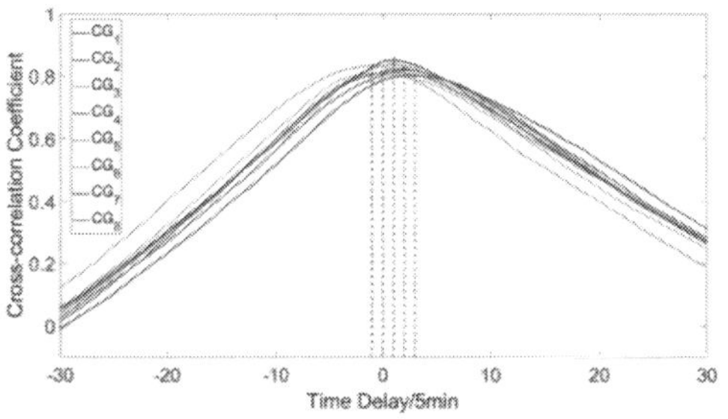

Fig. 4. Analysis Chart of the Average Time-delay Power Cross-correlation Coefficient.

Table 1. The ratio of days where the difference between the two correlation coefficients is less than 0.1.

CG1	CG2	CG3	CG4	CG5	CG6	CG8
0.808	0.838	0.798	0.869	0.788	0.869	0.818

We can know that in the majority of cases, the difference between the two correlation coefficients is less than 0.1. Therefore, we can use the power curves of the CP stations near DG_0 to calculate the *time-delay power cross-correlation coefficient*.

2.3 Data reconstructing method

According to the analysis of characteristics about the power curves of DP and CP, we propose a DP power data reconstructing method based on the electrical energy data of DP and CP, the power curves of CP, the spatial location of the stations and the correlation between the power curves of DP and those of CP.

In addition, traditional DP power data reconstructing method is using the power curves of nearest CP station, which would bring the great accuracy sometimes, such as in the sunny days. However, overall speaking, this model would bring the large variance. Therefore, we lead the idea of bagging in ensemble learning into the data reconstructing method and use several near-centralized photovoltaic power curves to reconstruct the missing distributed photovoltaic power data.

The specific data reconstruct process is as shown in Fig. 5. There are 9 steps in this method, of which the first two steps screen out target DP and CP stations: DG_t, CG_{tt1}, CG_{tt2} and CG_{tt3}. And other steps involve some calculation formulas, which are described in detail below.

Step4, step5 and step6 are in the cycle. We would introduce the situation that i is equal to one, to which other situations are similar.

In step4, we calculate the best delay time, dt_1 and dt_2, at the biggest time-delay power cross-correlation coefficients between the power curves of CG_{tt1} and those of CG_{tt11} and CG_{tt12} by (7). In the formula, CG_{tt11} is CG_{tt2} and CG_{tt12} is CG_{tt3}. $P_{CG_{tt1}}(t)$ and $P_{CG_{tt1j}}(t)$ respectively represent the power curves of the CP station CG_{tt1} and CG_{tt1j} on the day day_t. The explanations of the other variables in (7) are mentioned in (5).

$$\begin{aligned} &\arg\max_{dt_j} \, |corr(P_{CG_{tt1}}(t), P'_{CG_{tt1j}}(t))| \\ &s.t. \quad P'_{CG_{tt1j}}(t) = P_{CG_{tt1j}}(t+dt_j), j=1 \ or \ 2 \\ &\quad dt_j \in \{x \,|\, -x_{\max} \le x \le x_{\max}, x \in Z, x_{\max} \ge 0\} \end{aligned} \tag{7}$$

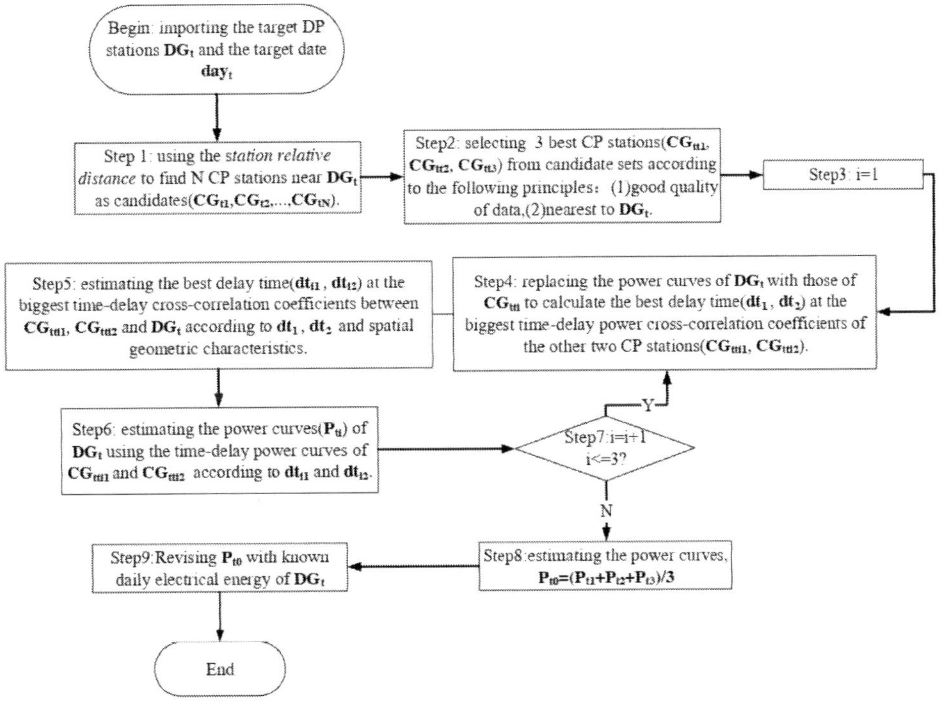

Fig. 5. Flow Chart of Data Reconstructing Method.

In step5, we use dt_1, dt_2 and spatial geometric characteristics to calculate the best delay time, dt_{11} and dt_{12}, at the biggest time-delay cross-correlation coefficients between CG_{tt11}, CG_{tt12} and DG_t. The spatial geometric characteristics of stations CG_{tt1}, CG_{tt11} and CG_{tt12} are shown in Fig. 6. We can calculate dt_{11} and dt_{12} by (8). The meanings of variables in the formula are shown in Fig. 6.

$$dt_{1j} = dt_j \times \cos(\theta_j) \times (1 - \frac{l_0}{l_j \times \cos(\theta_j)}), j=1 \ or \ 2 \tag{8}$$

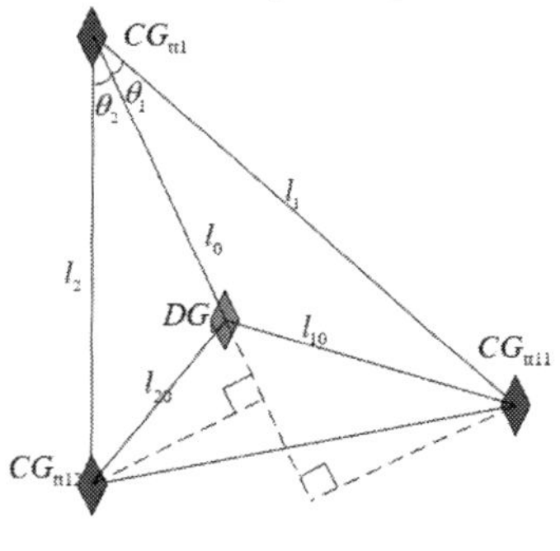

Fig. 6. The Spatial Geometric Characteristics.

In step6, we estimate the power curves P_{t1} of station DG_t by (9). In the formula, $P_{tt1}(t+dt_{11})$ and $P_{tt2}(t+dt_{12})$ respectively represent the power curves after the time translation of stations CG_{tt11} and CG_{tt12}. Q_t, Q_{t11} and Q_{t12} respectively represent the electrical energy of stations DG_t, CG_{tt11} and CG_{tt12} on the day day_t. The meanings of other variables in the formula are shown in Fig. 6.

$$\begin{aligned} P_{t1}(t) = &\frac{l_{20} \times Q_t}{(l_{10}+l_{20}) \times Q_{t11}} P_{tt1}(t+dt_{11}) \\ &+ \frac{l_{10} \times Q_t}{(l_{10}+l_{20}) \times Q_{t12}} P_{tt2}(t+dt_{12}) \end{aligned} \tag{9}$$

We can get P_{t1}, P_{t2} and P_{t3} at the end of the cycle. In step8, we can calculate the power curve of station DG_t by (10).

$$P_{t0} = (P_{t1} + P_{t2} + P_{t3}) / 3 \tag{10}$$

In step9, we revise P_{t0} with known daily electrical energy of DG_t by (11). In the formula, Q_{t0} represent the daily electrical energy of DG_t calculated by P_{t0}.

$$P_t = P_{t0} - P_{t0} \times (Q_{t0} - Q_t) / Q_{t0} \tag{11}$$

2.4 Evaluation indicators of data reconstructing effect

We can't get the real power curves of DP stations whose power data is large-scale missing, so we can't evaluate the data reconstructing effect in the actual situations. However, in order to evaluate the data reparing method proposed by us, we assume that we don't know the power curves of the DP station and use the data reparing method mentioned before to get the power curves. Then we use the actual power curves and the reparing power curves to calculate some indicators to evaluate the data reconstructing method, including *daily mean absolute error*, *daily mean relative error* and *daily mean capacity relative error*.

Daily mean absolute error(DMAE) can be calculated by (12). In the formula, the M represents total number of data points in a power curve. The P_{act} and P_{rep} respectively represent the actual power curve and the reparing power curve.

$$DMAE = \frac{1}{M}\sum_{i=1}^{M}|P_{act}(i) - P_{rep}(i)| \tag{12}$$

Daily mean relative error(DMRE) can be calculated by (13).

$$DMRE = \frac{1}{M}\sum_{i=1}^{M}\frac{|P_{act}(i) - P_{rep}(i)|}{P_{act}(i)} \tag{13}$$

Daily mean capacity relative error(DMCRE) can be calculated by (14). In the formula, S_{DG} represents the capacity of the DP station.

$$DMCRE = \frac{1}{M}\sum_{i=1}^{M}\frac{|P_{act}(i) - P_{rep}(i)|}{S_{DG}} \tag{14}$$

3 Case study

3.1 Simulation data

The simulation data originates from 8 CP stations and a DP station in Hebei Province, China, including their capacity, daily power curves, daily electrical energy and geographical position. It should be noted that the power curves of the DP station are only used for the evaluation of the data reconstructing but not for the reconstructing. Their basic information is shown in Fig. 1.

3.2 Simulation analysis

In order to evaluate the effect of data reconstructing, we propose some evaluation indicators and design two cases, whose explanations are following.

Case1: only using the power curves of the nearest CP station to reconstruct the power curves of the DP station according the capacity conversion.

Case2: using the data reconstructing method proposed in this paper based on the temporal and spatial characteristics between the power curves of the CP stations and those of the DP station.

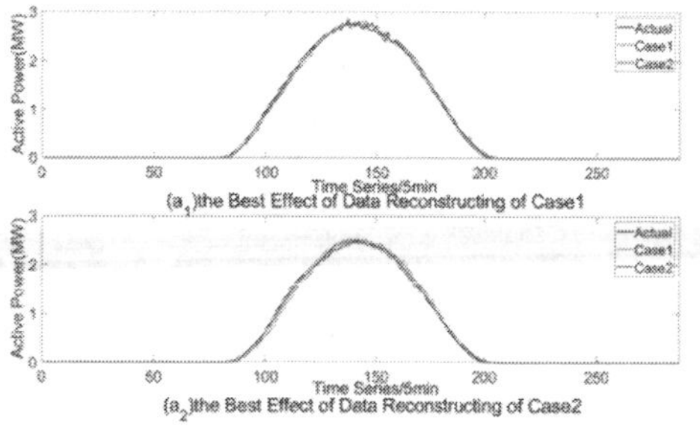

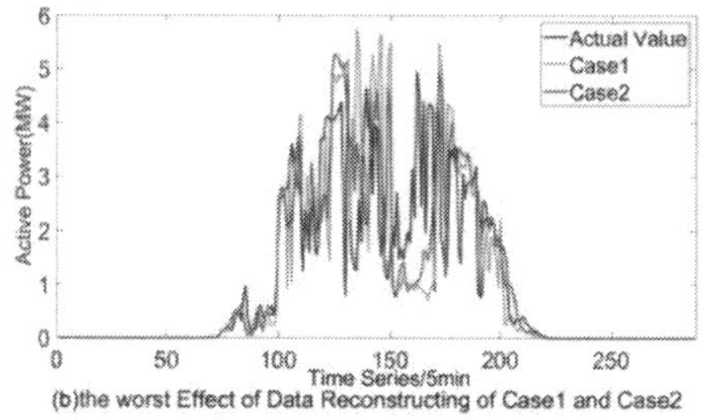

Fig. 7. The Best and Worst Results of Case1 and Case2.

Fig. 7 shows the simulation results of case1 and case2. Fig. 7(a1) shows the best effect of data reconstructing of case1 and Fig. 7(a2) shows the best effect of data reconstructing of case2.

We can know that case 1 and case2 can get the great effect of data reconstructing no matter in the best situation of case1 or in the best situation of case2. But actually, case2 can get the better effect of data reconstructing according to the evaluation indicators, which are shown in Table 2 and Table 3.

Table 2. The evaluation indicators of case1 and case2 when case1 gets the best effect of data reconstructing.

Indicators	DMAE	DMRE	DMCRE
Case1	0.0488MW	9.62%	0.81%
Case2	0.0426MW	6.46%	0.71%

Table 3. The evaluation indicators of case1 and case2 when case2 gets the best effect of data reconstructing.

Indicators	DMAE	DMRE	DMCRE
Case1	0.0771MW	11.94%	1.28%
Case2	0.0401MW	6.09%	0.67%

Fig. 7(b) shows the worst effect of data reconstructing of case1 and case2. Case1 and case2 get the worst effect of data reconstructing in the same day. Specifically speaking, case2 can get the better effect of data reconstructing according to the evaluation indicators, which are shown in Table 4.

Table 4. The evaluation indicators of case1 and case2 in the worst simulation results.

Indicators	DMAE	DMRE	DMCRE
Case1	0.9616MW	61.37%	16.03%
Case2	0.8017MW	53.64%	13.36%

We compares the DMCRE of case2 with those of case1 in different days in Fig. 8. We can know that case2 can get smaller DMCRE and show the better effect of data reconstructing in most of days.

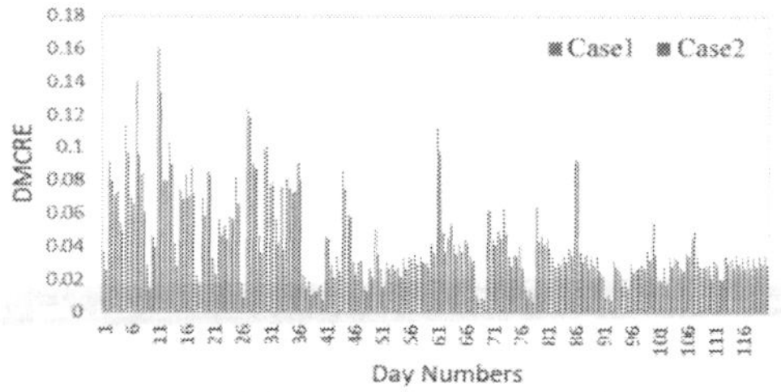

Fig. 8. The DMCRE of Case1 and Case2 in different days.

4 Conclusion

This paper defines some characteristic indicators to describe the temporal and spatial characteristics between the power curves of CP stations and those of DP stations, and proposes DP data reconstructing method based on the temporal and spatial characteristics. Finally, we verify the effectiveness of the proposed method by simulation based on the real photovoltaic power data. We can get the following conclusions:

(1) Converted to the same capacity, the daily electrical energy of CP is slightly larger than that of DP in most of cases.

(2) There is the correlation and the time-delay correlation between the power curves of DP and those of CP, which are related to the station relative distance.

(3) Compared with the traditional method, the proposed method can reduce the errors of data reconstructing effectively.

This work was supported by the project, named Research on Power Prediction Technology of Distributed Generation Cluster in the Power Grid of Jibei (SGJB0000TKJS1900140).

References

1. Y Wang, Q Chen, T Hong, et al. Review of Smart Meter Data Analytics: Applications, Methodologies, and Challenges. IEEE Transactions on Smart Grid(2018).
2. Y Lei, Z Li, Z Lu, etc. Review on the Research of Distributed Generation Technology and Its Impacts on Electric Power System. Southern Power System Technology, **5**, 46. (in Chinese, 2011)
3. J H Angelim ,C M Affonso. Impact of distributed generation technology and location on power system voltage stability. IEEE Latin America Transactions, **14**, 1758.(2016)
4. X Shen, M Cao. Research on the Influence of Distributed Power Grid for Distribution Network. Transactions of China Electrotechnical Society, **30**, 346. (in Chinese, 2015)
5. H Wang, L Ge, H Li, F Chi. A Review on Characteristic Analysis and Prediction Method of Distributed PV. Electric Power Construction, **38**, 1. (in Chinese, 2017)
6. Y Gong, Z Lu, Y Qiao, Q Wang. An Overview of Photovoltaic Energy System Output Forecasting Technology. Automation of Electric Power Systems, **40**, 140. (in Chinese, 2016)
7. S Gong, T Pan, D Wu, Z Ji. Research on missing data imputation of Mcro-Grid PV system based on MCMC. Renewable Energy Resources, **36**, 346. (in Chinese, 2018)
8. Zarzo M, Martí P. Modeling the variability of solar radiation data among weather stations by means of principal components analysis. Applied Energy, **88**, 2775.(2018)
9. R Yu, N Chen, N Miao, D Dang. A Repair Method for PV Power Station Output Data Considering Weather and Spatial Correlations. Power System Technology, **41**, 2229. (in Chinese, 2017)
10. B Lu, Q Fang, G Wang. Methods for identification and restoration of abnormal power data in distributed household rooftop photovoltaic device. Electric Drive Automation, **40**, 1. (in Chinese, 2018)

Capacity Configuration of Battery Energy Storage System for Photovoltaic Generation System Considering the High Charge-rate

Jiaming Li[1,*], *Ying* Qiao[1], *Guojing* Liu[2], and *Zongxiang* Lu[1]

[1]State Key Lab of Control and Simulation of Power Systems and Generation Equipments, Dept. of Electrical Engineering, Tsinghua University, Beijing 100084, China
[2]Economic Institute of Technology, State Grid Jiangsu Electric Power Company, Nanjing 210005, China

Abstract. Battery energy storage system (BESS) is one of the important solutions to improve the accommodation of large-scale grid connected photovoltaic (PV) generation and increase its operation economy. However, the strong intra-day volatility and severe curtailment of PV power sets a high demand of BESS charge-rate that is a key factor in operation models but ignored in current planning researches. This paper proposes a BESS capacity configuration model for PV generation systems which takes BESS's ability to (dis)charge exceeds its rated power into account. The best charge-rate and power & energy capacity of BESS are optimized by particle swarm optimization (PSO) algorithm. Through an analysis of the annual output statistics of PV power station in the northwest of China, the results show that when considering the high charge-rate of BESS, the optimal BESS capacity configuration rises and comprehensive income of the PV-BESS system will increase.

1 Introduction

With the energy transformation strategy proposed , photovoltaic (PV) generation installed capacity is increasing all over the word, and PV is becoming an important energy source in the future clean and renewable power system [1-3]. However, the development of PV is facing many challenges, the inherent features of PV such as strong volatility and randomness cause the difficulty to accommodate. The output characteristic of PV depends on the solar power, often mismatches with load demand, which gives rise to severe energy curtailment.

The BESS can help to absorb the electric energy in the PV peak period and release the electric energy in the low period, thus reduces the waste energy. With the continuous decrease of the cost of BESS, the integrated planning of energy storage and PV plants is seemed to be an effective, economical and reliable method, and some related research works have been done.

[4] gives the operation strategy and optimal capacity of BESS in a PV system based on the spectrum analysis of PV output. In [5], the method of interval estimation is used to predict probabilistic error of PV output, and the expected value of confidence level is obtained as the power allocation of BESS. [6] takes various scheduling modes into consideration, establishes the function of battery loss, then calculates the optimal BESS capacity.

However, in the existing researches on the capacity configuration of BESS for PV system, the charge-rate characteristics of BESS is often set to the rated value. Thus, the potential capability of BESS to enhance accommodation is underestimated, which leads to the excessive investment of BESS. The charge-rate characteristic of BESS indicates that, many kinds of chemistry battery (such as graphite-LiFePO4, vanadium redox flow battery and so on) have the ability to charge or discharge at several times (often up to 5-10) of its rated power [7]. When BESS with good high-rate characteristics is charged or discharged at high rate, its energy conversion efficiency is almost the same [8], which shows the advantage of high-power conversion. The BESS with good high-rate characteristic is being developed and applied, so the high-rate (dis)charging ability of BESS needs to be considered in the configuration of BESS capacity for PV system.

In order to give the capacity configuration of BESS for PV system, an economic optimization model of PV-BESS system is established, and the high-rate characteristics of BESS is considered in the model. The main works of the paper: Section 2 expounds the system structure and profit mode of PV generation system with BESS. In section 3, the high (dis)charge rate characteristics of BESS is given. The optimization model is given in Section 4. Section 5 shows a case study of a PV-BESS station in the northwest of China, and conducts a sensitivity analysis of various factors to BESS optimal (dis)charge rate. In Section 6 the conclusions are given.

* Corresponding author: jm-li19@mails.tsinghua.edu.cn

2 The technical and economic characteristics of PV-BESS generation system

2.1 The Structure of PV-BESS generation system

Main structure of centralized PV-BESS generation system is shown in Fig. 1, and BESS system (including BESS and BESS converter) is in the dotted box. The PV generation can charge BESS and transmit electric energy to the grid through the PV converter. BESS can absorb the electric energy of PV when it is in the charging state through BESS converter, and transmits electric power to the grid when it works on the discharging state.

Since the charge/discharge state and input/output power of BESS can be controlled manually, its cooperation with the PV generation system can effectively slow down the fluctuation of equivalent PV-BESS power generation, which improves the schedulability of PV system, and promotes the consumption.

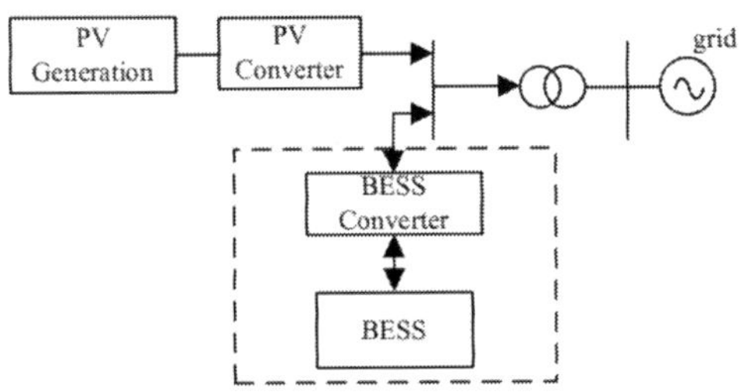

Fig. 1. Structure of PV-BESS system

2.2 Profit model of PV-BESS generation system under the policy of restraint

PV generation system gets profits from selling electricity to the grid. PV output power fluctuates strongly during the day, it has a strong power generation ability at noon on a sunny day. If all the electric power of PV is delivered to the grid at noon, normal grid dispatching will be affected seriously. Therefore, the power limitation measure is generally taken for PV system. The electric department formulates power limitation standards or PV plants, which means the power sold by PV system to the grid shall not exceed the power policy specified by the government department, the surplus power of PV will be wasted.

If additional BESS is set for PV system, BESS could absorb the surplus power of PV during the day time, then release electricity and sell it to the grid at night (Fig. 2). Thus, increase the income from electricity sales, however, the cost of BESS and BESS converter must be paid by the PV plants producer. In order to achieve better composite economic effects, it is necessary to allocate the capacity of BESS reasonably.

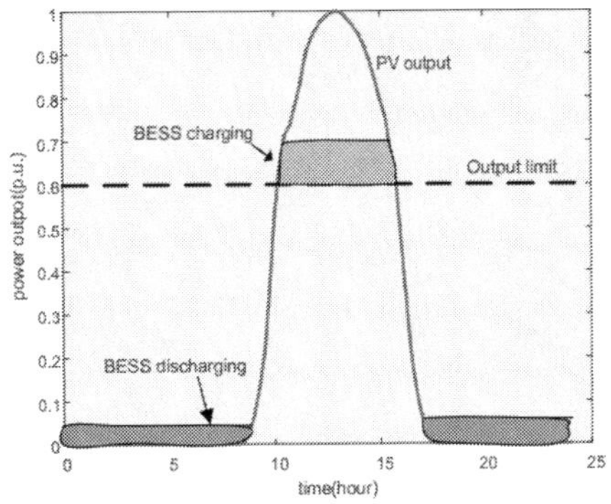

Fig. 2. Operation of PV-BESS system under the restraint policy

3 High-rate characteristics of BESS

Charge & discharge rate is the ratio of battery (dis)charge current to its rated capacity [9]. Generally, λ is used to represent the ratio of battery charge and discharge current. For a 1200mAh battery, λ=0.2 represents (dis)charging at 240mA (0.2 times of 1200mAh), and λ=1.0 represents 1200mA (1 time of 1200mAh). Under the condition of constant (dis)charge voltage, λ can also represent the ratio of charge and discharge power to its rated power.

Supposing that the actual battery capacity won't change with its (dis)charging rate, the relationship between the continuous discharge time of a full-charged BESS and the discharging rate is shown in Fig. 3. (in the case of λ=1, the discharging time is 60min).

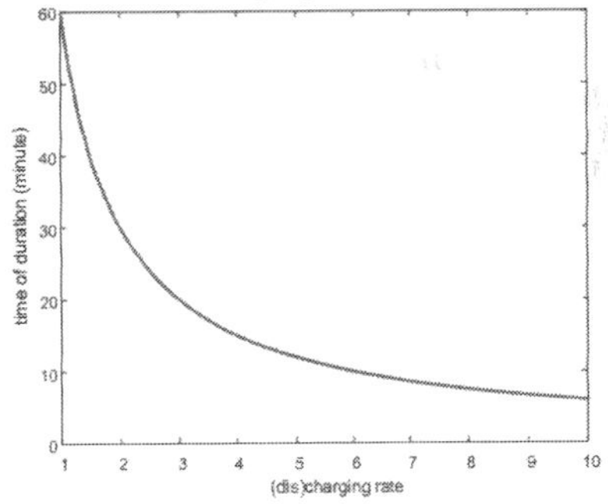

Fig. 3. The relationship between time duration and (dis)charging rate of BESS

Considering the energy conversion efficiency of BESS, the actual capacity that can be released by BESS is slightly smaller than in the ideal case, that is, the actual curve will be below the curve drew in .For most types of BESS, the larger the (dis)charge rate, the smaller capacity can be discharged or charged by BESS.

Using the data from the literature [10], the actual capacity of a particular lithium iron phosphate battery under different (dis)charge ratio is shown in Fig. 4, the capacity of energy that is released by the BESS under $\lambda = 2$ shows 90% of the nominal capacity.

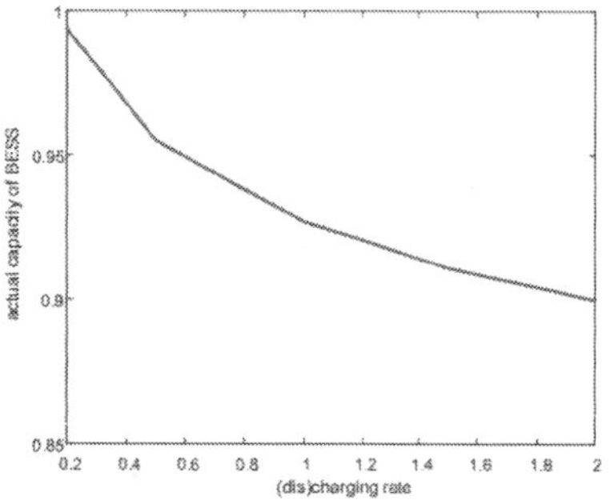

Fig. 4. The relationship between actual capacity and (dis)charging rate of BESS

4 Optimization model

The optimization model is built on the objection function of the extra annul income by setting BESS in a PV generation system. Because of the nonlinear nature of the optimization model, particle swarm optimization (PSO) algorithm is used to solve the problem.

4.1 Optimization function

4.1.1 Optimization objective

Taking the maximum extra annual income of PV generation system provided by the setting of BESS as the optimization target. The extra annual income includes the increased annual generation income and the construction & operation costs of BESS (equivalent to one year), and it can be expressed as followed:

$$\max\{I_{np} = (I_{grid} - I_{BESS})\} \tag{1}$$

Where I_{np} is the extra annual income of PV generation provided by BESS (dollars/year), I_{grid} is the extra annual power generation income of PV generation (dollars/year), I_{BESS} is the annual construction & operation costs of BESS (dollars/year).

4.1.2 Constraint condition

The power & capacity cost of BESS and the BESS converter cost matched with the highest (dis)charge rate ($\lambda_{\max}$) are required to be paid for the construction of BESS system. I_{BESS} can be expressed as:

$$I_{BESS} = (C_P P_{BESS} + C_E E_{BESS} + \lambda_{\max} C_{cov} P_{BESS}) / n_{BESS} \tag{2}$$

Where C_P is sum construction and operation cost of unit BESS rated power (dollars/MW), C_E is sum construction and operation cost of unit BESS rated energy capacity (dollars/MWh), C_{cov} is sum construction and operation cost of unit BESS converter rated power (dollars/MW), P_{BESS} is rated power of BESS (MW), E_{BESS} is rated energy capacity of BESS (MWh), $\lambda_{\max}$ is maximum charge/discharge rate of BESS, n_{BESS} is life span of BESS (year)

Increased revenue from electricity sales due to the construction of BESS can be expressed as:

$$I_{grid} = C_{grid} E_{grid} \tag{3}$$

Where C_{grid} is electricity sale income of PV system (dollars/MWh), E_{grid} is extra electricity sales capacity due to the establishment of BESS (MWh/year).

The BESS adopts the operation mode of charging once a day and discharging once a day to sell the electric energy stored in the daytime at night to the grid. Assuming that the electric power sold to the grid is limited to P_l, the electric output capacity of PV system on the i *th* day at time *t* is $P^i_{solar}(t)$. If the rated power of BESS is set as P_{BESS}, the maximum charge/discharge rate is set as $\lambda_{\max}$, supposing the configured energy storage capacity is large enough, the maximum energy that can be absorbed by BESS on the daytime is:

$$E^i = \int_{P^i_{solar}(t) > P_l} (\min\{P^i_{solar}(t), P_l + \lambda_{\max} P_{BESS}\} - P_l) dt \tag{4}$$

Where E^i is maximum energy can be absorbed on the *i th* day (MWh), P_l is power limit (maximum power sent to the grid) of PV-BESS system (MW), $P^i_{solar}(t)$ is output capacity of PV system on the *i th* day at the time of *t* (MW).

Considering the capacity limitation of BESS, as well as the charging & discharging efficiency and the state of charge (SOC) limitation, the increased electricity sales capacity on this day due to the construction of BESS are as follows:

$$E^i_{grid} = \eta_{ch}\eta_{dis}(SOC_{\max} - SOC_{\min})\min\{E^i, E_{BESS}\} \tag{5}$$

The annual extra electricity sales capacity can be expressed as:

$$E_{grid} = \sum_{i=1d}^{i=365d} E^i_{grid} \tag{6}$$

4.2 Solving algorithm

Because the objective function has nonlinear term, particle swarm optimization (PSO) algorithm is used. The core of PSO algorithm is to make use of the individual information sharing in the group, so that the movement of the whole group can generate a process from disorder to order in the solution space, so as to obtain the optimal solution of the problem. PSO algorithm is suitable for solving nonlinear mathematical programming problems without analytic expressions. Let the capacity of BESS (E_{BESS}) and the maximum (dis)charge rate set for BESS be the variables, and the solving flow chart is shown in Fig. 5.

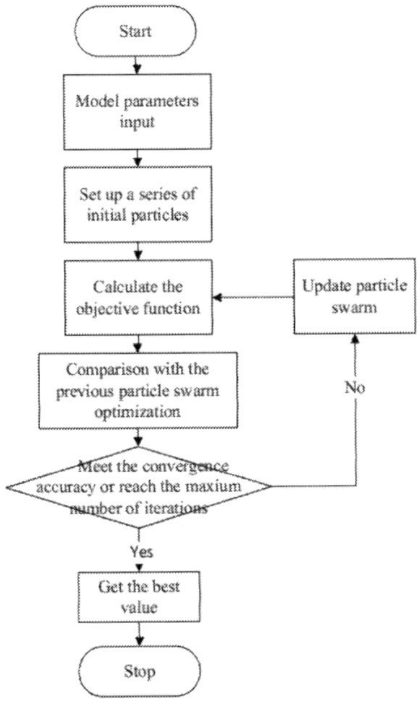

Fig. 5. Solving flow chart

5 Case study

5.1 Parameters of the case

Based on the annual actual power generation data (sampling interval is 15min, 35040 sample points totally) of a 100MW PV power station in Inner Mongolia, the above-mentioned maximum annual income model is used for simulation calculation, and the optimal capacity of energy storage device is compared under different maximum charge-rate of energy storage, and the sensitivity analysis of power station revenue is carried out. The BESS considered in this paper is lithium-ion battery, and the simulation parameters are set as Table 1:

Table 1. Value of parameters

Parameter	Value	Parameter	Value
P_l	50MW	SOC_{max} SOC_{min}	0.9 0.1
P_{BESS}	5MW	η_{ch}, η_{dis}	0.9
E_{BESS}	0-100MWh	C_P	1*10^5\$/MW
λ_{max}	1-5	C_E	1.5*10^5\$/MWh
n_{BESS}	15years	C_{cov}	1*10^5\$/MW
C_{grid}	90\$/MWh		

5.2 Result analysis

If taking both λ_{max} and E_{BESS} as the optimization variables, the result of the optimization is λ_{max} =2.6, E_{BESS} =54.2MWh, and the extra annual income is I_{np} =11.6(10^4\$). If λ_{max} is fixed in the optimization model, Fig. 6 shows the variation trend of optimal BESS capacity & income with different λ_{max}. The optimal capacity allocation of BESS increases with the increase of the maximum (dis)charge rate, and finally tends to be stable. Meanwhile, the benefits of BESS for PV system increase first and then decrease, and achieve the maximum at λ_{max} =2.6. Compared with the case of the high-rate (dis)charge characteristics is ignored (λ_{max} =1, E_{BESS} =24.1MWh, I_{np} =6.0*10^4\$), the total annual income increases 93.3%.

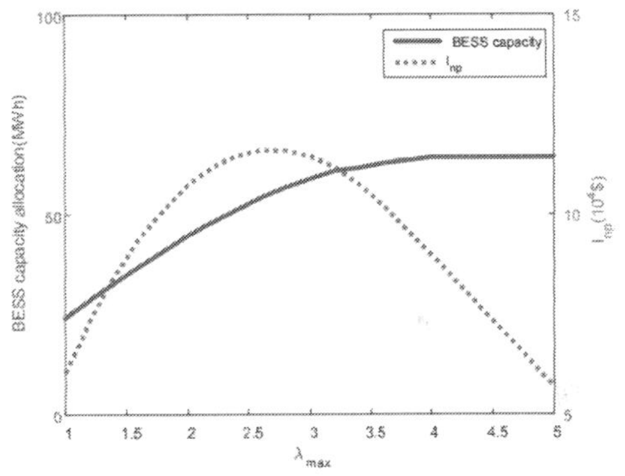

Fig. 6. Variation trend of optimal BESS capacity & income with different λ_{max}

5.3 Sensitivity analysis

The government's policy of limiting PV power sent to the grid affects the space in which BESS can play a role in increasing the generating capacity of PV stations, and the cost of BESS affects the economy of its installation and operation. Therefore, P_l and C_E are the two key factors that influence the optimization results. Fig. 7, Fig. 8 separately express the variation trend of optimal λ_{max} with different P_l and C_E, when P_l declines or C_E is reduced, the optimal λ_{max} rises. PV stations with severe power limit needs higher charge rate, and with the cost reduction of BESS, battery with higher (dis)charge rate is more economical.

https://doi.org/10.1051/e3sconf/202018203003

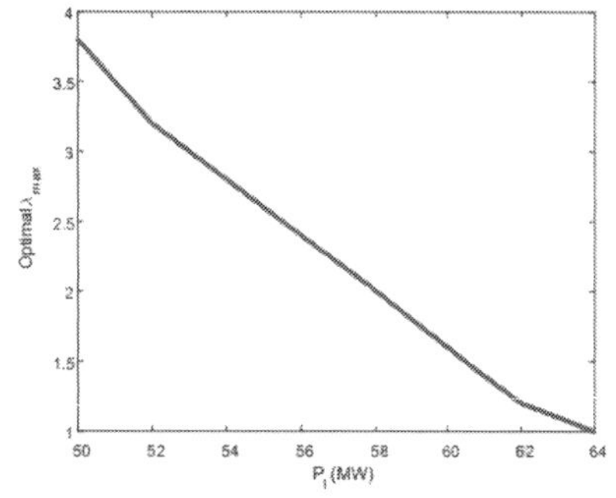

Fig. 7. Variation trend of optimal $\lambda_{\max}$ with different P_l

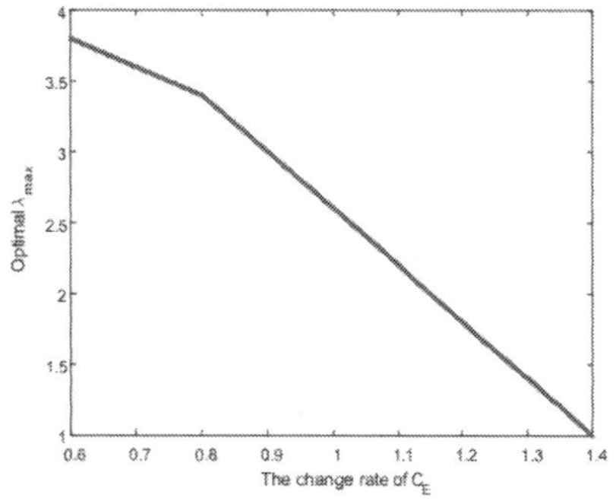

Fig. 8. Variation trend of optimal $\lambda_{\max}$ with different C_E

6 Conclusion

This study set out to calculate the most economical capacity of BESS for a PV generation considering the influence of (dis)charge-rate, where an optimization model is established. The results show that with considering the characteristics of high (dis)charge rate of BESS, both the BESS configuration capacity and comprehensive income of PV generation system will increase, especially in the PV station with severe curtailment. Besides, with trend of BESS cost decreasing, the optimal charge rate will rise, with which PV generation system will receive higher benefits.

This work is supported by the State Grid Technology Project "Research and Application of Large-scale Energy Storage Planning Technology for Received-end Power Grid". (5102-201918309A-0-0-00)

Reference

1. ZHAO Bo, ZHANG Xuesong, HONG Bowen. Energy penetration of large-scale distributed photovoltaic sources integrated into smart distribution network[J]. Electric Power Automation Equipment, 2012,32(8): 95-100.
2. CHEN Shuyong, BAO Hai, WU Chunyang, et al . Direct grid-tie power control method for distributed photovoltaic generation[J]. Proceedings of the CSEE, 2011, 30(10): 6-11.
3. TIAN Qi, ZHAO Zhengming, DENG Y, et al . Simulation and experimental study about reverse model of photovoltaic cells[J]. Proceedings of the CSEE, 2011, 31(23): 121-128.
4. CHEN Gang, YUAN Yue, FU Zhixin. Application of storage battery to restrain the photovoltaic power fluctuation[J]. Proceedings of the CSU-EPSA, 2014, 26(2): 27-31.
5. LIN Shaobo, HAN Minxiao, ZHAO Guopeng, et al. Capacity allocation of energy storage in distributed photovoltaic power system based on stochastic prediction error[J]. Proceedings of the CSEE, 2013,33(4): 25-33.
6. XU Yan, HE Chen, FU Chao, et al. Analysis of energy storage capacity allocation considering optimal economy of PV & Battery cogeneration system under different scheduling modes[J]. Acta Energiae Solaris Sinica,2019,40(6): 1632-1640.
7. John Wang, Ping Liu, Jocelyn Hicks-Garner, et al. Cycle-life model for graphite-LiFePO4 cells[J]. Journal of Power Sources,196(2011):3942-3948.
8. LI Shujuan, LI Xinran, HUANG Jiyuan, et al. Capacity Configuration of Lithium Titanate Battery in Whole Grid Frequency Regulation Based on Rate Characteristic[J]. High Voltage Engineering,44(1):145-151.
9. Aghamohammadi M R, Abdolahinia H. A new approach for optimal sizing of battery energy storage system for primary frequency control of islanded Microgrid. International Journal of Electrical Power & Energy Systems, 2014, 54(01):325-333.
10. TANG Jin XU Guofeng LI Jianling. Study on cycling performance of lithium iron phosphate battery at different discharge rates[J]. Nonferrous Metals Science and Engineering,8(5):95-102.

https://doi.org/10.1051/e3sconf/202018203004

Modeling study of ground source heat pump system based on the dynamic load of underground engineering

Jintian Li [1,*], *Yunzhe* Ji [1], *Bo* Wang[2], and *Ling* Xie [1]

[1] Navy Logistics College in PLA, Tianjin 300450, China
[2] Rocket Force University of Engineering in PLA, Xian 710025, China

Abstract. The load properties of underground engineering have an important influence on operating characteristics of ground source heat pump system. It has important reference value for design and operation management that Simulation analyzing operating conditions of ground source heat pump system under dynamic load conditions. It took an underground engineering as an example for dynamic load calculation in the paper, and simulated operating characteristics of ground source heat pump system under three operating conditions. The calculation results show that the engineering maintenance and management period is conducive to the recovery of soil temperature, and it improves the COP value of the unit. Some measures should be taken to restore soil temperature for long-term continuous operation of underground engineering. The use of heat recovery to make domestic hot water can relieve the problem of soil thermal imbalance to some extent. It is beneficial to improve heat pump unit performance.

1 Introduction

A large amount of waste heat is generated when equipment operation in underground engineering. The internal temperature of the engineering will rise, and it will lead to deterioration of the working environment, equipment operation failure, even system paralysis if not properly handled. At present central air-conditioning system were used in most of underground engineering in China. Because the engineering load in the spatial distribution is not uniform, it will cause huge energy waste. The use of ground heat exchanger can not only release the residual heat into the soil, but also overcome the problems of central air-conditioning system can't partition temperature control and energy consumption, it is very flexible and convenient. Vertical buried pipe heat exchange is less affected by outdoor temperature, has higher efficiency, and its application range is the most extensive. But the cost is higher, so it is important to establish accurate heat exchange model for the economic and sustainable operation of the ground source heat pump system [1,2]. The load characteristic has important influence to the operating characteristics of the ground source heat pump system [3,4,5]. The operating characteristics will also change when the load changes in underground engineering. Compared with the surface building, the underground engineering is affected by the surrounding geothermal heat, the internal temperature change is small. Because of the influence of day and night changes, and the method of use and other factors, the underground engineering load will change greatly. For example, when the engineering is running at full load, internal personnel and equipment will emit a lot of residual heat, and when the engineering is under maintenance, the rest of the heat is less, or even a thermal load. Therefore, it is of great practical significance to analyze the operation of the ground source heat pump system under dynamic load conditions [6,7].

2 Dynamic load analysis of underground engineering

Take an underground engineering hall as an example, the length l is 12.3m, the width b is 4.8m, the arch height h_1 is 2.4m, the arch height f is 1.1m, the rock material parameter is: λ= 2.04 W/ (m· K), a =0.0031 m^2/h, inner surface heat exchange factor h=7 W/(m^2·K). The natural temperature of the rock is 17.6°C, and the room is warmed up by 600h to reach the air temperature design requirement of 26°C.

The internal air system of the hall is analyzed, and its dynamic thermal balance equation can be expressed as:

$$V_a(\rho c)_a \frac{\partial T_a(\tau)}{\partial \tau} = Q_W(\tau) + Q_R(\tau) + Q_S(\tau) + Q_X(\tau) - Q_{JZ}(\tau) \quad (1)$$

$T_a(\tau)$: indoor air temperature, °C;

$Q_W(\tau)$: heat transfer quantities through enclosure, W;

$Q_R(\tau)$: personnel heat release, W;

$Q_S(\tau)$: equipment heat release, W;

$Q_X(\tau)$: the heat of introduced or permeation fresh air, W;

$Q_{JZ}(\tau)$: the heat carried away by ground source heat pump unit, W。

* Corresponding author: ljt_1008@126.com

https://doi.org/10.1051/e3sconf/202018203004

Next, the heat is analyzed for each part.

2.1 Heat storage in the enclosure

For the limited length arch section project, the calculation can be approximated to the equivalent sphere, and the heat transfer of its internal enclosure structure can be calculated by press [8].

$$Q_w = F \cdot h(t_a - t_\infty)\left[1 - f_2(Fo, Bi)\right]m \tag{2}$$

$f_2(Fo,Bi) = \frac{Bi}{1+Bi}(1 - e^{F_0(1+Bi)^2} erfc[(Bi+1)\sqrt{Fo}]$, $F_0 = a(\tau + \tau_0) / R_0^2$, $Bi = \alpha R_o / \lambda$, In the equation, τ_0 is the project warm-up period time, here take 600h; F is the inner surface area of the enclosure, m²。

2.2 The heat release by personnel

The amount of heat dissipated by personnel can be calculated in a press:

$$Q_r = 0.96 \cdot n \cdot q \tag{3}$$

In the equation, q *is the* total heat dissipation, W; n is the number of personnel, depending on the mode of the engineering, the number of personnel and the length of stay as shown in Table 1.

Table 1. Number of Hall personnel in different engineering modes.

the mode of the engineering	Full-load operation		Maintenance management period	
the length of stay	Day (12h)	Night (12h)	Day (12h)	Night (12h)
the number of personnel	30	15	5	0

According to the literature [9,10] , the amount of thermal emission of personnel in underground engineering is 52W at 26°C and the amount of latent heat emission is 73W.

2.3 Equipment and lighting heat dissipation

The heat dissipation in the hall mainly comes from the operation of various equipment, the thermal load is 150 to 400 W/m². For easy analysis, the total load of equipment and lighting is 15kW, and the maintenance management period is 10% of the total load.

2.4 fresh air load

Assuming that the amount of new air in the hall varies with the number of people in the room, the new air is usually calculated by introducing the interior air supplement of the project, which can be calculated according to the following formula:

$$Q_X = nL_p c_p \rho'_{out}(T'_{out} - T_a) / 3600 \tag{4}$$

When fresh air is introduced from outside, due to the influence of the surrounding structure, the new wind outside the project will generally be reduced by about 3 to 5°C, in this article Δt=2°C, The new wind introduction is 25m³/h, and the maintenance management period does not take into account the new air load.

$$T'_{out} = T_{out} - \Delta t \tag{5}$$

Among the equation, T_{out} is outdoor time-by-time dry ball temperature in summer, $T_{out'}$ is the inner temperature of the engineering.

3 The establishment of the model of the ground source heat pump system

The ground source heat pump system is modeled using TRNSYS software. Its greatest feature is the open and modular structure design, the user has direct access to the source code and part model of the software, and can be self-edited and added to the actual needs of the part model [11,12]. The model of the ground source heat pump system built in this paper mainly includes the buried pipe heat exchanger, heat pump unit, building load coupling model, temperature control model, load input and output model, as shown in Fig. 1.

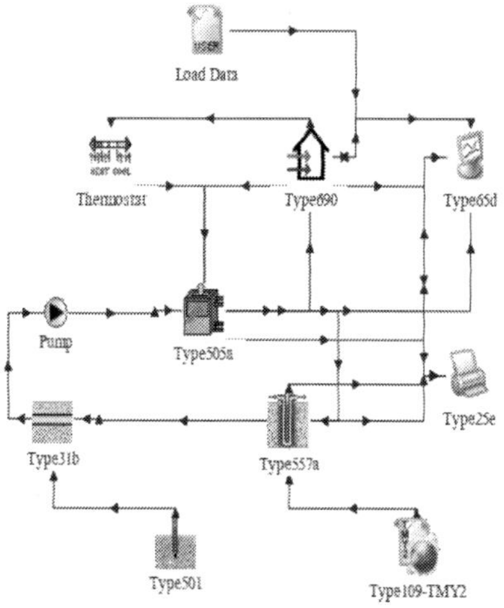

Fig. 1. TRNSYS simulation interface.

Type9: Data can be read from an external file and then joined to other parts of the simulation software. Because TRNSYS can't calculate the hot and humid load of underground buildings well, this paper uses MATLAB to calculate the hot and humid load of large residual heat room, imports the data into Type9, and carries on the subsequent simulation calculation.

Type108: Room temperature control signal. The room temperature is controlled by this module at 24°C. Thus indirectly control the operation of the heat pump unit;Type690: Buildings; Type557: Vertical buried pipe;

Type31: This module is used for horizontal tube temperature changes; Type3: Pump; Type505: Water-air water source heat pump; Type501: This module allows the original soil temperature to be pre-set; Type109: Weather parameters.

The relevant parameters such as the buried pipe used for the calculation are shown in Table 2.

Table 2. Simulation related parameters.

parameters	value	unit
Drilling diameter (d_b)	0.12	m
Branch Outer Diameter (d_{po})	0.032	m
Branch Inner Diameter (d_{pi})	0.027	m
Branch Spacing (x_c)	0.06	m
Heat conductivity of backfill materials (λ_g)	2.4	W/(m·K)
Tube wall thermal conductivity (λ_p)	0.6	W/(m·K)
Drilling depth (H)	100	m
Geothermal conductivity (λ_s)	2.4	W/(m·K)
Volume thermal capacity ($\rho_s c_s$)	2.0	MJ/(m³·k)
Circulating water flow (V)	1100	kg/h

4 Simulation results and analysis

4.1 Maintenance management period operation

The dynamic load of the underground engineering is analyzed Using the model in the oversection, and the in-and-out temperature and energy efficiency ratio of the unit are shown in Fig. 2 when the system operates for 20 years. It can be seen that the soil temperature can be better restored due to the maintenance period. The maximum incoming and the outgoing water temperatures of the units are 24.5℃ and 29.3℃ for the 10th opera Ting cycle. The unit COP is minimum 3.91, maximum 4.72, average 4.15.

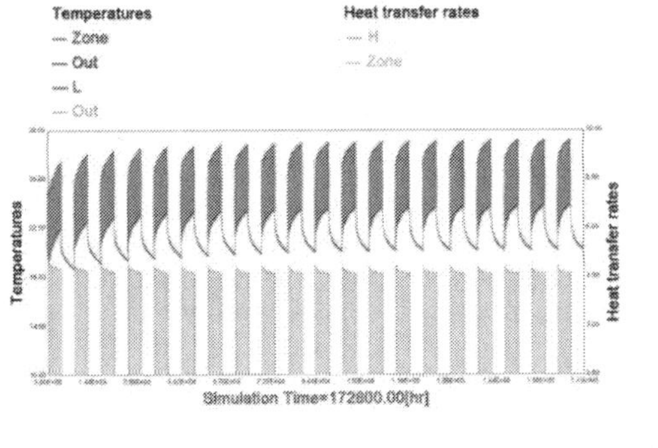

Fig. 2. Simulates the temperature and energy efficiency ratio of the incoming and outgoing units for 20 years of maintenance.

4.2 Continuous full load operation

In order to compare the impact of maintenance management period on the operation of the ground source heat pump system, it is assumed that the engineering has been running at full load, and simulates the water temperature and unit energy efficiency ratio of the ground source heat pump system for 10 years of continuous operation. The results are shown in Fig. 3. The maximum water inlet temperature in the 10th year rises to 27℃, and the maximum water temperature rises to 31.8℃. COP fell from 4.68 to 3.68, with an average COP of 3.89. It can be seen that the unit will release more heat to the soil for 20 years of intermittent operation, but due to intermittent operation of the soil temperature to a certain extent to restore, compared with continuous operation, the running effect of unit intermittent operation is better.

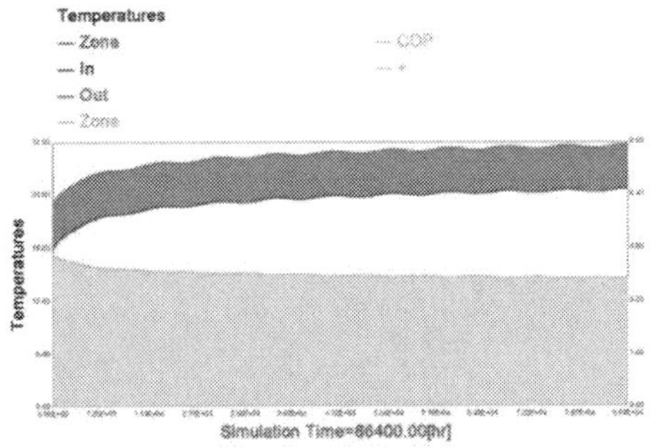

Fig. 3. Simulates the in-and-out water temperature and COP of the unit for 10 years of continuous operation at full load.

4.3 Condensation heat recovery method

Using the condensation heat generated by the heat pump unit to collect domestic hot water is an economical and effective way to solve the problem of soil heat imbalance. It is convenient and energy-saving to make domestic hot water by means of condensation heat recovery. Assuming that the number of personnel in the project is 30, the domestic hot water is based on 50L/d standard. Hot water needs to be heated from 20℃ to 55℃, and the daily water supply time is 6:00-22:00. Because the room is constant temperature control, the heat pump unit operation is intermittent. When the heat pump unit stops working, the domestic hot water is heated by the auxiliary electric heater. The ground-source heat pump system using heat recovery to make domestic hot water is shown in Fig. 4.

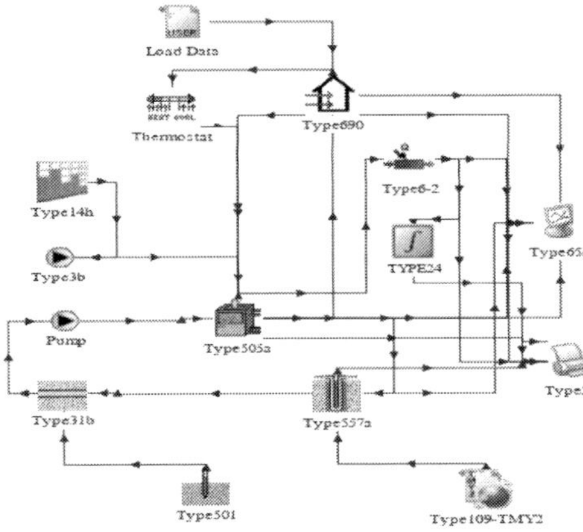

Fig. 4. Heat recovery method for making live hot water TRNSYS simulation.

The room temperature, the temperature of the buried pipe in and out of water, the average temperature of the rock and earth, the temperature of the hot water supply are analyzed when the heat pump system operate for 48h. As can be seen from Fig. 5, the temperature in the large residual heat room can be controlled near 24°C, the temperature of the heat pump unit in and out of water after the production of domestic hot water is significantly reduced, which is conducive to solving the problem of soil heat imbalance, but also improve the COP of the unit.

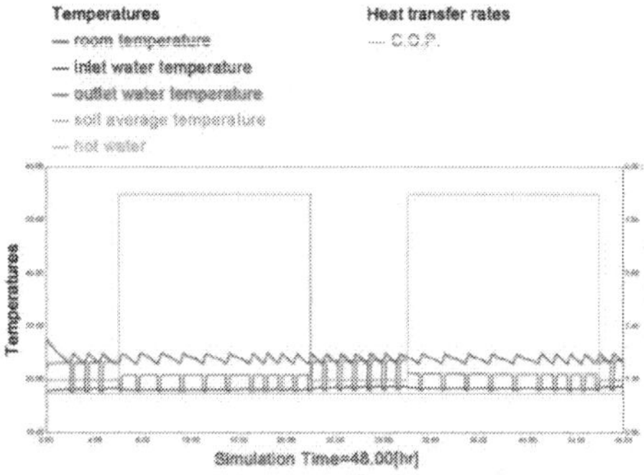

Fig. 5. The results of hot recovery mode running for 48h

The water temperature of the buried pipe was simulated for 10 years of continuous operation. As can be seen from Fig.6, before and after the heat recovery was used, the water temperature of the buried pipe was reduced from 26.95°C to 23.78°C. Ten-year auxiliary heating power consumption is 22,700 million J, and if all the domestic hot water is heated by electric heating, the required electricity consumption is:

$$Q = cm\Delta t = 7.92 \times 10^{11} J \quad (6)$$

Therefore, the use of heat recovery method to make domestic hot water can not only save electricity but also to a certain extent to solve the problem of the thermal and hot load imbalance of the buried pipe, thereby improving the operating performance of the ground source heat pump system.

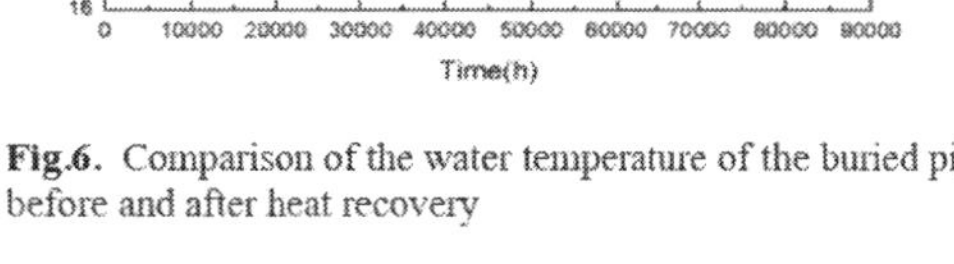

Fig.6. Comparison of the water temperature of the buried pipe before and after heat recovery

4 Conclusions

When the underground engineering adopts the ground source heat pump air conditioning system, different operating conditions will have an impact on the operating characteristics of the system, and the operation mode with engineering maintenance management period is conducive to the recovery of soil temperature, and also has a catalytic effect on improving the COP of the unit. If the unit needs a long period of continuous operation, it should to take certain measures to restore soil temperature. The use of condensation heat generated to make domestic hot water, can alleviate the problem of soil heat imbalance to a certain extent, is conducive to improving the performance of the heat pump unit.

References

1. Zheng He-yi, Diao Nai-ren, Fang Zhao-hong. Analysis of heat transfer inside a borehole of vertical ground heat exchangers[J]. Solar Energy Journal, 2004, 25(3):399-405.
2. Wang Liang, Xu Yong-jun. Operating characteristics of ground source heat pump system in non-equilibrium cooling and heating load[J]. Journal of Thermal Science and Technology, 2016, 5(2):154-158.
3. Leszek R, Małgorzata H. Cumulated energy consumption in a heat pump system using a U-tube ground heat exchanger in a moderate climate[J]. Energy & Buildings,2015,9(6):118-127.
4. Dai L H, Shang Y, Li X L. Analysis on the transient heat transfer process inside and outside the borehole for a vertical U-tube ground heat exchanger under short-term heat storage[J]. Renewable Energy,2016,8(3):1121-1129.
5. Chen Jin-hua, Xia Lei, Daniel Mmereki. Simulation and experimental analysis of optimal buried depth of the vertical U-tube ground heat exchanger for a ground-coupled heat pump system[J]. Renewable Energy,2015,7(3):46-54.
6. Zhang Jian-song. Experimental Study on Heat Transfer Mechanism of Ground Source Heat Pump Based Single U-Type Vertical Buried Pipe Inlet and Outlet Spacing[C]. Advanced Science and Industry Research Center. Proceedings of 2018 International Conference on Energy, Power, Electrical and Environmental Engineering, 2018:35-40.
7. Maria L F, Alberto L, José A. Analysis of the thermal interference between ground source heat pump systems in dense neighborhoods[J]. Science and Technology for the Built Environment,2019,25(8):1069-1080.
8. Mao Jin-feng, Han Xu. The theory and application of thermal and humidity in underground engineering [M]. Beijing: China Construction Industry Press, 2009.
9. Underground Building HVAC Design Manual [M]. China Construction Industry Press, 1983.
10. Liu Jun, Yao Yang, Wang qing-Qin. Simulation and analysis of heat transfer in underground building enclosure structure [C]// National HVAC Refrigeration 2004 Academic Collection.

11. Hou Gao-yang, Taherian H, Li Long-jun. A predictive TRNSYS model for long-term operation of a hybrid ground source heat pump system with innovative horizontal buried pipe type[J]. Renewable Energy, 2019.
12. Liu Zhi-jian, Li Yuan-wei, Xu Wei. Performance and feasibility study of hybrid ground source heat pump system assisted with cooling tower for one office building based on one Shanghai case[J]. Energy, 2019,1(3):28-37.

Modeling of Photovoltaic MPPT Lead Acid Battery Charge Controller for Standalone System Applications

Rodney H.G. Tan[1,*], *Chee Kang* Er[1], and *Sunil G.* Solanki[1]

[1]UCSI University, Faculty of Engineering, Technology and Built Environment, 56000 Kuala Lumpur, Malaysia.

Abstract. This paper presents the circuitry modeling of the solar photovoltaic MPPT lead-acid battery charge controller for the standalone system in MATLAB/Simulink environment. A buck topology is utilized as a DC-DC converter for the charge controller implementation. The maximum power of the photovoltaic panel is tracked by the Perturb and Observe MPPT algorithm. The battery charge controller charges the lead-acid battery using a three-stage charging strategy. The three charging stages include the MPPT bulk charge, constant voltage absorption charge, and float charge stage. The performance analysis of the model is carried out in the following aspects, there are MPPT tracking performance, battery charging performance and overall charge controller efficiency performance are benchmarked with commercial MPPT charge controller for validation. The performance result shows that the MPPT is capable to track to the PV panel maximum point at any solar irradiance variation within 0.5 seconds with maximum power tracking efficiency up to 99.9 %. The three-stage charging strategy also successfully demonstrated. The overall charge controller average efficiency achieved up to 98.3 % which matches many high end commercial solar PV MPPT charge controller product specifications. This validated model contributes to a better sizing of PV panel and battery energy storage for the small and medium standalone PV system.

1 Introduction

Solar photovoltaic energy has gained significant attention in the past decade. It is one of the fastest growing renewable energy up to 181 GW installation globally as of 2018 [1]. This is simply due to the steady reduction of photovoltaic module costs over the past decade. Moreover, the ease of installation, flexible scalability and less maintenance due to no moving part involved in its operation make it widely accepted by many industries. The photovoltaic system relies on sunlight to generate electricity. This limits its application to only day time when sunlight is available. To utilize solar energy at any time of the day including when sunlight is not available, battery energy storage is one of the solutions. Therefore solar PV charge controller plays an important role to make this solution feasible.

The solar PV charge controller is widely used in standalone system applications including street lighting [2], telecommunication base station, rural electrification [3], etc. A Solar PV MPPT charge controller includes an MPPT tracker and a battery charge controller. The MPPT tracks the maximum power from the PV panel and delivers it to the battery charge controller. The charge controller charges the battery through a multi-stage charging strategy to effectively charge the battery without damaging the battery caused by excessive charge gassing and overheat. Many past works of literature are covering various MPPT algorithms review [4-7], MPPT modeling and implementation in Simulink [8]. However, these works of literature do not evaluate the MPPT tracking performance in terms of tracking time and tracking efficiency. There is also no discussion on how this MPPT can be integrated with a battery charge controller. On the battery charge controller side, there are literatures solely present on multi-stage charging strategies [9], comparative study of various multi-stage charger [10], solar PV charge controller [11]. However, these charge controller presented are charged without MPPT and there is no performance analysis in terms of charger efficiency. There are also literatures presented on the modeling of solar PV MPPT charge controller [12-13], however the presented model lack of modeling detail and does not have any performance analysis in term of efficiency and no benchmark with commercial charge controller for model validation. In summary, the above literatures presented models that are lack of completeness and didn't present MPPT and overall charge controller efficiency performance analysis and benchmark with commercial MPPT charge controller for model validation. This paper presents the complete modeling of the solar PV MPPT charge controller in Simulink and its performance analysis in the following sections.

2 Methodology

The overview of the solar photovoltaic MPPT battery charge controller model developed in MATLAB/Simulink environment is shown in Fig. 1. It consists of a solar PV array, DC-DC converter, battery,

* Corresponding author: rodneytan72@hotmail.com

and MPPT charge controller block. Inside the MPPT charge controller block consists of a Perturb & Observe MPPT algorithm and a lead-acid battery three-stage charge controller is shown in Fig. 2. The MPPT charge controller block includes a P&O MPPT tracker and a lead-acid battery three-stage charger. The MPPT charge controller block output a PWM control signal to switch the switching device of the DC-DC converter. This is a common design for many commercial solar PV MPPT battery charge controller. The solar PV array model and battery model are directly obtained from the Simulink Simscape Electrical blockset library. The model is capable of charging a 48 V battery from 2 kW PV array source. This model is tested and simulated under the Simulink environment for performance analysis. The following sections explain the circuitry model and charge controller block in detail.

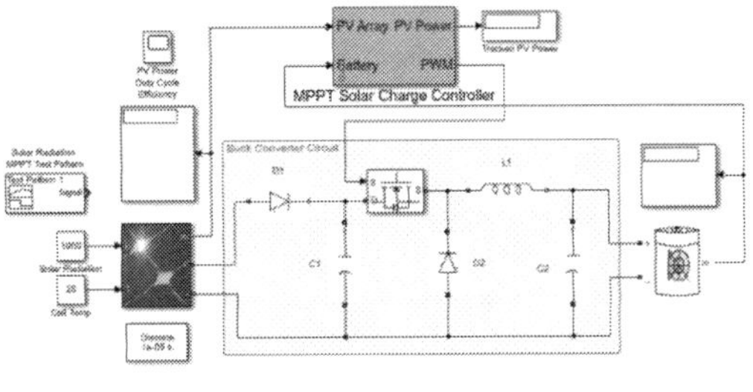

Fig. 1. Overview of solar PV MPPT charge controller model.

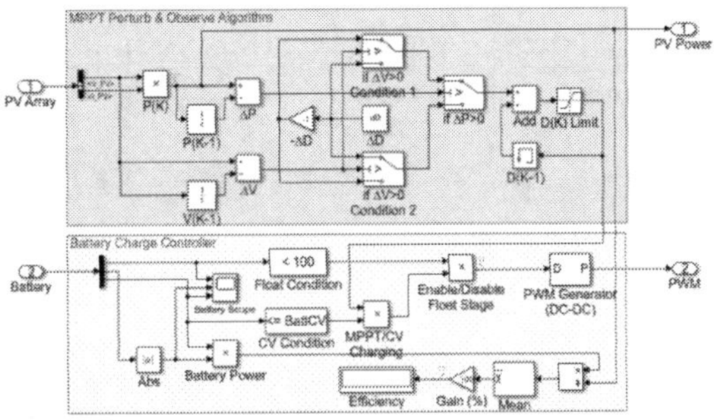

Fig. 2. Subsystem of MPPT charge controller block.

2.1. DC-DC Converter

The DC-DC converter converts the DC input voltage source to either higher or lower DC output voltage. Since the PV array voltage is higher than the battery voltage, a buck topology is commonly chosen for solar PV charge controller application [14-16]. The buck converter operates as a regulator to step down the input voltage from the PV array while maintaining its power delivery to charge the battery. This is achieved by stepping down the input voltage and increasing the output current delivered to the battery. The buck converter circuit consists of a MOSFET switching device, a high power inductor, Schottky diode, and an input and output capacitor as shown in Fig. 1. The reverse blocking diode D1 is used to prevent the reverse flow of current back to the PV array from the battery during night time. The MOSFET R_{on} is set to 0.02 Ω and the Schottky diode D2 forward voltage is set to 0.5 V. The MOSFET is switch by a pulse generator with 1000 Hz switching frequency.

The output voltage of the buck converter can be determined by the ratio between V_{out} the output voltage and V_{in} the input voltage of buck converter (1) where D is the duty cycle of the PWM signal.

$$D = \frac{V_{out}}{V_{in}} \tag{1}$$

The duty cycle can also determine the effective input resistance seen by the PV array source where MPPT can be achieved. The effective input resistance R_{in} seen by the source can be determined in (2) where R_{load} is the load or battery resistance and D is the duty cycle.

$$R_{in} = \frac{R_{load}}{D} \tag{2}$$

The buck converter inductor ripple current peak to peak magnitude under steady-state condition can be determined in (3) where V_{in} is the input voltage from the PV array, D is the duty cycle, f_{sw} is the switching frequency and L is the inductor value.

$$\Delta I_L = \frac{V_{in}\,\mathrm{D}(1-D)}{f_{sw}L} \tag{3}$$

Similarly, the buck converter output capacitor ripple voltage peak to peak magnitude under steady-state condition can be determined in (4) where V_{in} is the input voltage from the PV array, D is the duty cycle, f_{sw} is the switching frequency and C is the capacitor value.

$$\Delta V_C = \frac{V_{in}\,\mathrm{D}(1-D)}{8Lf_{sw}^2C} \tag{4}$$

From equation (1), (3) and (4), for inductor value of 10 mH and capacitor value of 1000 uF, the buck converter ripple current and voltage can be determined. For switching frequency f_{sw} of 1000 Hz, PV input voltage V_{in} of 120 V, duty cycle D of 0.4, the inductor ripple current ΔI_L is around 0.288 mA and output capacitor ripple voltage ΔV_c is around 2.3 nV, which are small enough to be a flat DC.

2.2 Maximum Power Point Tracking Algorithm

The P&O MPPT is commonly used in many small and medium commercial solar PV charge controller and grid-connected inverter due to its tracking effectiveness and simplicity of implementation. The MPPT algorithm track the maximum power of the PV array and output its duty cycle relevant to the tracked maximum power to the battery charge controller. The P&O MPPT algorithm flowchart is shown in Fig. 3. This algorithm operates

 https://doi.org/10.1051/e3sconf/202018203005

based on the trial and error process in tracking the maximum power point. The algorithm observes the power changes and perturbs the PV panel operating voltage by changing the duty cycle to the converter switching device which in turn changing the effective input resistance (2) of the buck converter. It then observes again if it reaches the maximum power and this process repeats itself indefinitely.

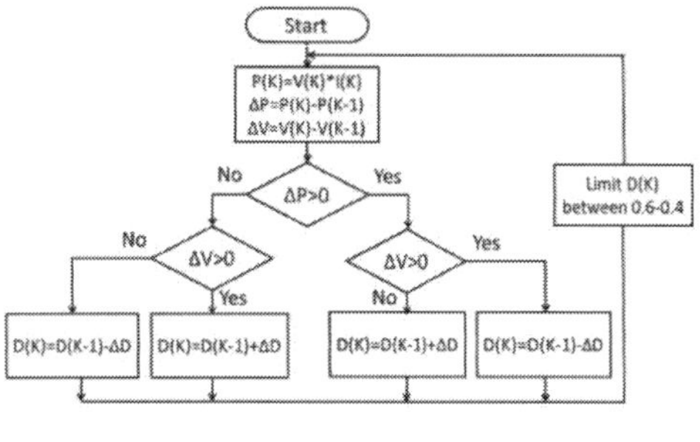

Fig. 3. Perturb & Observe MPPT algorithm flowchart.

The implementation of the MPPT Perturb & Observe Algorithm in Simulink is shown in Fig. 4. It is implemented using only Simulink blocks without using any scripting code. Each block is labelled with its function with respect to the flowchart. The P&O MPPT algorithm takes in voltage and current reading from the PV array, the previous sample (K-1) function is carried out by the unit delay block. The three if-else conditions of the P&O algorithm are carryout by the condition switch block, the ΔD block allows the user to set the perturbation step size of the duty cycle, the duty cycle increment and decrement function are carried out by an adder with a memory block D(K-1) feedback loop. The D(K) limit block limit the duty cycle exceeding the range between 0.4 to 0.6. The output of the duty cycle is connected to the battery charge controller section. The PV power output is also connected to the battery charge controller for conversion efficiency computation.

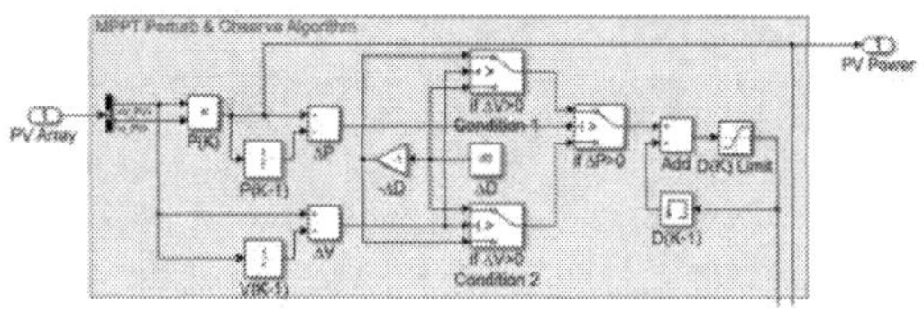

Fig. 4. Perturb & Observe MPPT algorithm implementation in simulink.

2.3 Lead Acid Battery Charger Controller

The battery charge controller was developed to charge a lead-acid battery using the three-stage charging method. The three-stage charging includes the constant current charging, constant voltage charging, and float charging stage. The first stage constant current charging also refer as bulk charging stage where the charge current is charge at its rated capacity, in this case, the charge current is at MPPT. The second stage constant voltage charging also refer as absorption charging stage where the battery is charged at a constant voltage, in this stage the MPPT is disabled. Finally, the third stage float charge simply maintains the battery State of Charge (SoC) at 100 % when the battery is fully charged. This is to prevent the battery gassing reaction and overheat due to an uncontrolled excessive charge at over 100 %. The flow chart of the battery charge controller is shown in Fig. 5. The charge controller measures the battery SoC and voltage. In the first condition, if the battery SoC is less than 100 % then the charger enters into the constant current or constant voltage charging stage, else enters float stage where duty cycle D(K) is zero. In the second condition, the battery voltage level is used to determine the MPPT bulk or constant voltage absorption charging stage. If the battery voltage is less than the constant voltage set point, the charger switch to MPPT constant current bulk charging stage, else disable MPPT and switch to constant voltage absorption charging stage.

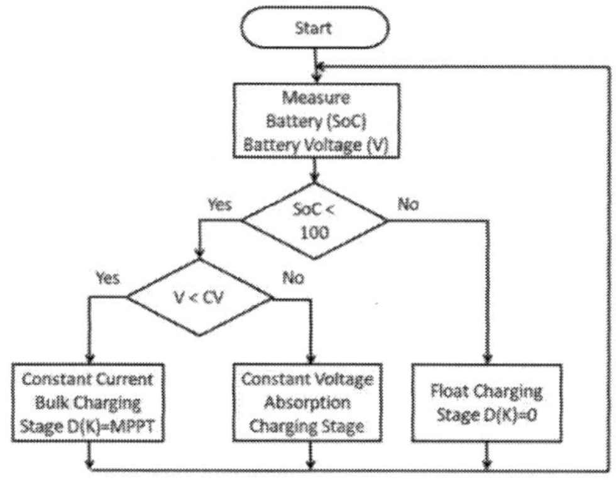

Fig. 5. Three-stage lead acid battery charge controller flowchart.

The implementation of a battery charge controller in Simulink is shown in Fig. 6. The battery charge controller read in the battery voltage and SoC as input. The Simulink utilized the compare to constant block as if condition to determine if the battery SoC less than 100 %, if the condition is true it will disable the float stage by allowing the MPPT duty cycle to pass through the multiply block so that the charger enters into either bulk or absorption charging stage. If the condition is false, meaning the battery already reaches its SoC of 100 % it will enable float stage by sending zero to set the duty cycle to zero.

The condition to determine the charger to enter MPPT or constant voltage charging stage is carryout by the battery voltage is less than or equal the constant voltage set-point, if the condition is true then enable MPPT charging by allowing MPPT duty cycle to pass through the MPPT/CV charging multiply block, then through the disable float stage multiply block to the PWM generator block for output to the buck converter switching device. If the condition is false, the charger enters into constant voltage absorption charging stage by rapid switching between MPPT and zero through the MPPT/CV multiply block so that the battery voltage can maintain at the constant voltage set-point until float stage is reached.

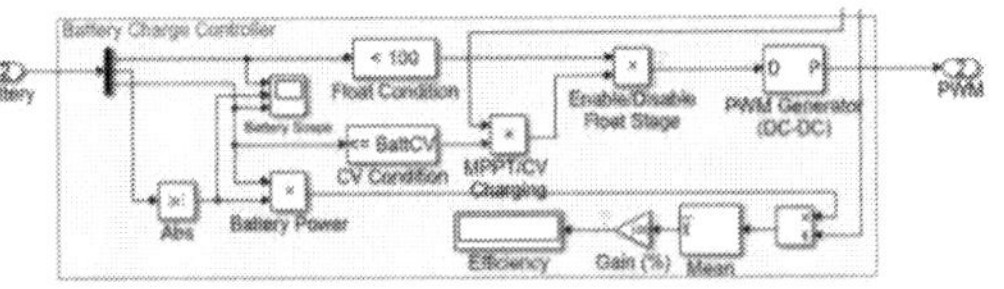

Fig. 6. Lead acid battery charge controller implementation in simulink.

3 Results and Discussion

The MPPT battery charge controller for the standalone photovoltaic system model has been simulated successfully in MATLAB/Simulink environment for performance analysis. The Simulink model is configured using ode23tb (Stiff/TR-BDF2) solver with a variable step. The discrete type of simulation is set with a sample time of 10 μs per sample. The following sections detail out the performance analysis of the model in four aspects, there are MPPT tracking performance, battery charging performance, overall efficiency performance and validation with commercial MPPT charge controller.

3.1. MPPT Tracking Algorithm Performance

The tracking ability of the MPPT algorithm is verified through this performance analysis. The performance of the MPPT tracker is carryout by varying the solar irradiance input to the PV array ranging from 100 W/m^2 to 1000 W/m^2 in 100 W/m^2 step with 2 seconds interval. At 1000 W/m^2 a sudden drop to 100 W/m^2 is also introduced at 20s of the step irradiance to simulate fast passing cloud. The solar irradiance pattern above is constructed using a signal building block. The PV array is set based on a commercial module with maximum power voltage of 30.9 V, maximum power current of 8.1 A, open-circuit voltage of 36.6 V and short-circuit current of 8.75 A . It configures in 2 parallel strings with 4 panels per string and produces a total of 2 kW power. The module temperature is set at 25°C. The MPPT perturbation step ΔD is set to 1e-6. To run this model without any algebraic loop error, the algebraic loop solver has set to Line Search based algorithm in Simulink. The performance of the MPPT tracker is shown in Fig. 7.

It can be seen that the MPPT algorithm has successfully tracked close to its maximum power from the PV array with its respective solar irradiance level. The tracking time of the MPPT is less than 0.5 seconds with the highest tracking efficiency of 99.9 % at solar irradiance of 300 W/m^2 with a tracked maximum power of 579 W. The inefficiency is due to the tracking error caused by the P&O algorithm oscillate near its maximum point. This is a known inherent problem of the P&O MPPT algorithm. The tracking error can be further reduced by reducing the perturbation step ΔD, however, this will increase the tracking time proportionally.

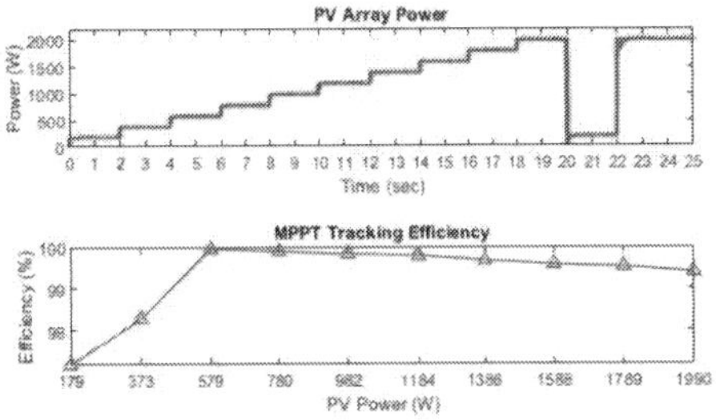

Fig. 7. MPPT P&O algorithm tracking performance.

3.2 Battery Charge Controller Performance

The performance of the battery charge controller is carryout by charging the battery through the sequence of MPPT bulk, constant voltage absorption and float charging stages to demonstrate the three stages charging capability of the charger. The battery block is set to a lead-acid battery, 48 V, 100 AH capacity, state of charge of 99.7 % and simulation duration of 60 sec. The PV array solar irradiance is set at 1000 W/m2 which produces 2000 W of power. The charge controller PWM frequency is set at 1000 Hz and the battery constant voltage set point is set at 55.4 V. The performance of the battery charge controller is shown in Fig. 8.

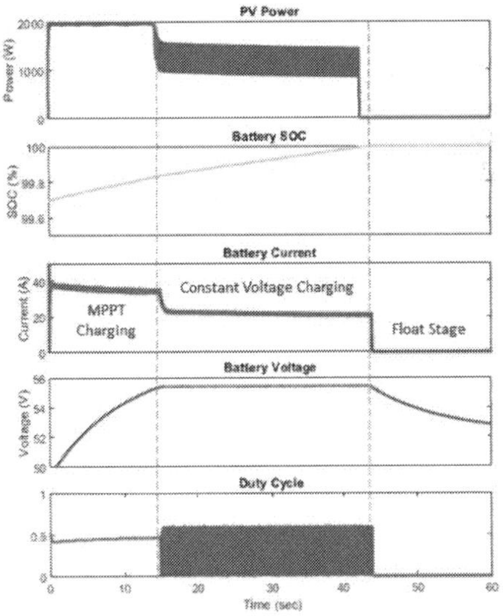

Fig. 8. Lead acid battery charge controller performance.

It can be seen that initially, the charge controller charges the battery at MPPT bulk charging stage when the battery SoC and voltage are less than 100 % and 55.4 V respectively. The charger switch to a constant voltage absorption charging stage when the battery voltage reaches 55.4 V at 15 sec. In this stage, the charge controller no longer charges the battery at MPPT power, the duty cycle is switch between MPPT and zero to maintain the constant voltage of 55.4 V which can be seen at duty cycle time from 15 sec to 42 sec. When the SoC reaches 100 % at 42 sec, the charge controller switch to float stage where the duty cycle is zero and the battery voltage drop to the floating voltage of 52.8 V.

3.3 Overall Charge Controller Efficiency Performance

The overall efficiency performance of the MPPT battery charge controller model is carryout by running the charge controller charging the battery under various solar irradiance ranging from 100 W/m^2 to 1000 W/m^2 at 100 W/m^2 interval in Simulink environment. The efficiency can be determined by the ratio of the output charging power delivery to the battery to the input power from PV array. The efficiency computation is implemented in the battery charge controller shown in Fig. 5. The mean block is used to average out the efficiency with a window size of 1 sec. This is to smooth out the efficiency oscillation caused by the MPPT algorithm. The efficiency performance is shown in Fig. 9. The average conversion efficiency up to 2000 W range is about 97.9% with the highest efficiency of 98.4 % and the lowest efficiency of 97.4%. The efficiency performance of this model is benchmarked with a commercial MPPT Charger controller from EPEVER iTracer IT6415ND as reference. The model overall efficiency performance within the efficiency range of 98% to 97% of the commercial charge controller product specification. The model conversion efficiency can be further improved by utilizing a lower drain-source resistance MOSFET and a lower forward voltage diode.

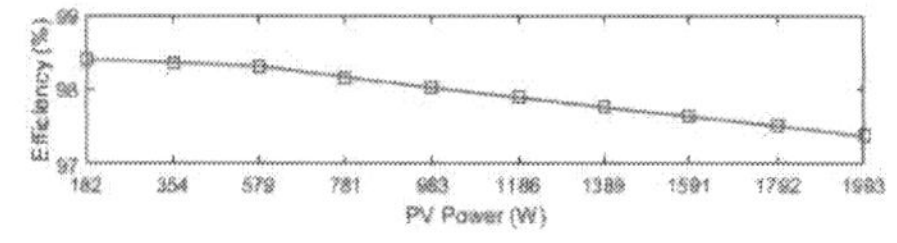

Fig. 9. Overall battery charge controller performance.

3.4 Model Performance Validation with Commercial MPPT Charge Controller

To validate the performance of this model is comparable with the commercial solar PV MPPT charge controller. A commercial solar PV MPPT charge controller is set up with 2 kW PV panel to charge 48 V battery. The commercial solar PV MPPT charge controller used for validation is from EPEVER iTracer-IT6415ND with MPPT tracking efficiency of 99.5% overall charge controller efficiency of 97-98%. The 48 V battery source consists of eight 12 V Lead Acid Gel type batteries with 100AH capacity. Four 12 V battery connected in series to form one set of 100 AH battery. And two sets of 48 V 100 AH battery set are connected in parallel to form a 200AH capacity battery system. The complete validation experiment setup is shown in Fig. 10.

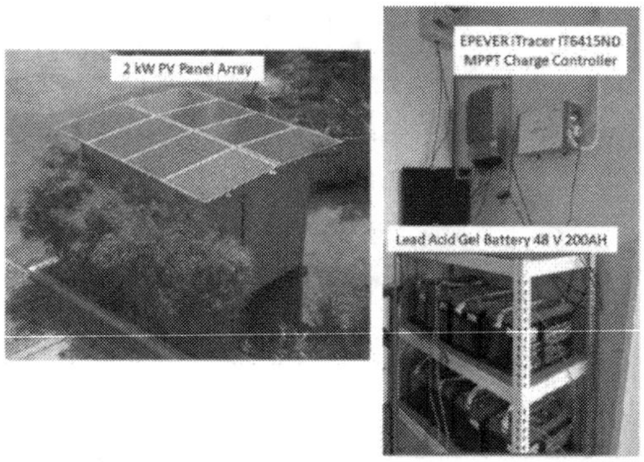

Fig. 10. Commercial MPPT charge controller experiment setup.

The validation experiment begins with recording the commercial MPPT charge controller PV input power, charging battery power and battery state of charge at 1 min interval. The recorded data is then saved as reference for the Simulink model validation. The simulated data is then compared with the recorded data for validation and performance comparison. Fig. 11 shows the validation results of PV input power, overall conversion efficiency and battery state of charge for 30 mins. It can be seen that the PV power input of the model and commercial MPPT charge controller almost overlapped with each other which indicates the model simulation PV input power is the same as the commercial PV MPPT charge controller data. The average overall efficiency performance of the PV MPPT charge controller model developed in Simulink is 98.3% which is comparable with the commercial PV MPPT charge controller recorded data with an average efficiency above 98.1%. The Simulink model has slightly higher efficiency, it could be the Simulink model environment is more ideal or can possibly cause by difference topology and components tolerance used in the commercial MPPT charge controller.

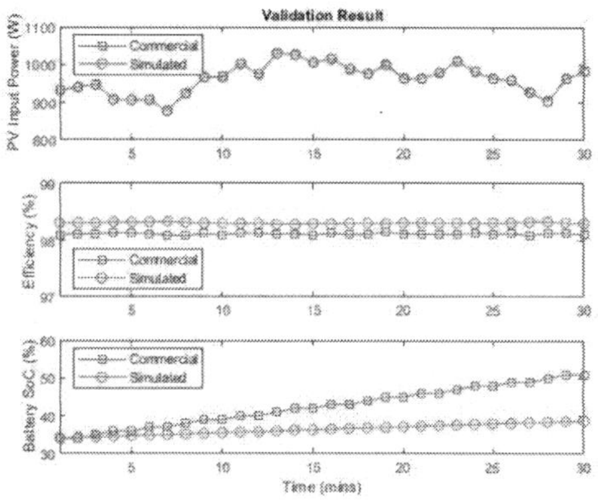

Fig. 11. Commercial and model simulation results.

The Battery State of Charge (SoC) charging rate also comparable with the real battery with about 12% of SoC difference. The difference could possibly cause by the following factors. The difference Lead Acid battery model parameters in Simulink and the real Lead Acid Gel

battery. The battery SoC computation in the Simulink model is based on the integration of current flowing in with respect to time and its capacity, where else battery SoC computation for the commercial charge controller is based on the battery voltage level. The battery model in Simulink is always simulated under new battery conditions without being affected by the aging and capacity derated condition. On the other hand, the experiment setup battery used for validation has been 10 months old. The aging battery can cause the battery capacity to derate which possibly causes the SoC charging rate to be higher than the Simulink model. The MPPT charge controller Simulink model presented in this paper is fully reproducible, with that in mind the model in MATLAB/Simulink presented in this paper is made available by the authors for the reader to download at Mathworks official MATLAB Central File Exchange link below
https://www.mathworks.com/matlabcentral/fileexchange/73115-mppt-solar-charge-controller-model

4 Conclusion

A detail circuitry modeling of a Solar PV MPPT battery charge controller model in Simulink is presented. The MPPT P&O tracking algorithm, buck converter circuit and three stages charge controller are clearly explained and are fully reproducible. The MPPT battery charge controller is capable to charge a 48 V lead-acid battery through tracking the maximum power from the 2 kW PV array power source and regulate the charging using a three-stage charging strategy. It achieved an average overall efficiency of 98.3% which matches many high end commercial solar PV MPPT charge controller product specifications. The Simulink model presented can be flexibly changed to meet any commercial MPPT charge controller with similar topology. The performance of the Simulink model also validated with a real commercial solar PV MPPT charge controller experimental setup. This validated model contributes to a better sizing of PV panel and battery energy storage for small and medium standalone PV systems.

The authors would like to acknowledge the PSIF grant Proj-In-FETBE-040 and conference funding support CONF-INT-FETBE-261 from UCSI University research management unit CERVIE so that the research and publication of this work are made possible.

References

1. Renewables 2019 Global Status Report, Renewables Now (REN21), 40 (2019)
2. B. N. Mohapatra, A. Dash, B. P. Jarika, Power Saving Solar Street Lights, International Journal of Emerging Technologies in Engineering Research, **5**, 105-109 (2017)
3. Noridzuan Idris, Ahmad Maliki Omar, Sulaiman Shaari, Stand-Alone Photovoltaic Power System Applications in Malaysia, *The 4th International Power Engineering and Optimization Conference*, (2010).
4. Trishan Esram, Patrick L. Chapman, Comparison of Photovoltaic Array Maximum Power Point Tracking Techniques, IEEE Transactions on Energy Conversion, **22**, 439-449 (2007).
5. Mohamed A. Eltawil, Zhengming Zhao, MPPT Techniques for Photovoltaic Applications, Renewable and Sustainable Energy Reviews, **25**, 793-813 (2013).
6. Nur Atharah Kamarzaman, Chee Wei Tan, A Comprehensive Review of Maximum Power Point Tracking Algorithms for Photovoltaic Systems, Renewable and Sustainable Energy Reviews, **37**, 585-598 (2014).
7. Nabil Karami, Nazih Moubayed, Rachid Outbib, General Review and Classification of Different MPPT Techniques, Renewable and Sustainable Energy Reviews, **68**, 1-18 (2017).
8. Mohamed A. Enany, Mohamed A. Farahat, Ahmed Nasr, Modelign and Evaluation of Main Maximum Power Point Tracking Algorithms for Photovoltaics Systems, Renewable and Sustainable Energy Reviews, **58**, 1578-1586 (2016).
9. E. Koutroulis, K. Kalaitzakis, Novel Battery Charging Regulation System for Photovoltaic Applications, *IEE Proceeding - Electric Power Applications*, **151**, 191-197 (2004).
10. Ankur Bhattacharjee, Design and Comparative Study of Three Photovoltaic Battery Charge Control Algorithms in MATLAB/Simulink Environment, International Journal of Advanced Computer Research, 2, 129-135 (2012).
11. M. Lokesh Reddy, P.J.R. Pavan Kumar, S. Aneel Manik Chandra, T. Sudhakar Babu, N. Raasekar, Comparative Study on Charge Controller Techniques for Solar PV System, *Energy Procedia*, **117**, 1070-1077 (2017).
12. B. Sree Manju, R. Ramaprabha, B.L. Mathur, Modelling and Control of Standalone Solar Photovoltaic Charging System, *International Conference on Emerging Trends in Electrical and Computer Technology* (2011).
13. Salman Salman, Xin Al, Zhouyang Wu, Design of a P&O Algorithm based MPPT Charge Controller for a Stand-alone 200W PV System, Protection and Control of Modern Power Systems, **3**, 1-8 (2018).
14. Byamakesh Nayak, Alivarani Mohapatra, Kanungo Barada Mohanty, Selection Criteria of DC-DC Converter and Control Variable for MPPT of PV System Utilized in Heating and Cooking Applications, Cogent Engineering, **4**, 1-16 (2017).
15. M.H Taghvaee, M.A.M. Radzi, S.M. Moosavain, Hashim Hizam, M. Hamiruce Marhaban, A Current and Future Study on non-isolated DC-DC Converters for Photovoltaic Applications. Renewable and Sustainable Energy Reviews, **17**, 216-227 (2013).

https://doi.org/10.1051/e3sconf/202018203005

16. Illan Glasner, Joseph Appelbaum, Advantage of Boost vs Buck Topology for Maximum Power Point Tracker in Photovoltaic Systems, *IEEE Proceedings of 19th Convention of Electrical and Electronics Engineers in Israel*, (1996)

E3S Web of Conferences **182**, 03006 (2020) https://doi.org/10.1051/e3sconf/202018203006
CPEEE 2020

Performance of heat pump system for water heating in European climate

Yauheni Kachalouski[1,*], *Matuška* Tomáš[2]

[1]Department of Environmental Engineering Faculty of Mechanical Engineering, Czech Technical University in Prague, Technická 4, 16607, Prague, Czech Republic
[2]University Centre for Energy Efficient Buildings, Czech Technical University in Prague, Třinecká 1024, 27343 Buštěhrad, Czech Republic

Abstract. A performance of air-to-water and brine-to-water heat pump for domestic hot water preparation was analyzed in different climates of Europe. Air-source heat pumps are widely used energy source for domestic applications with low operation costs. Their system *SPF* was found at level of 3.0 for domestic hot water preparation for medium latitudes in Europe. Investigations on the ground-source heat pump performance show their average *SPF* are close to air-source technology.

1 Introduction

Heat pumps are widely used as an ambient heat transformer for space heating (SH) and domestic hot water (DHW) preparation in Europe [1], however a lot of countries have only small share of installed heat pump power, e.g. such as Czech Republic. But general tendency in the market growth can be seen from statistics and reports [1]. Overall, the industry is optimistic for a continued market growth and more markets we return to growth in future. Moreover, according to the field measurements the overall performance of heat pumps in service is increasing [2]. This takes place in connection with new heat pump installations. It was investigated that on average the air-source heat pumps perform with a system performance factor (*SPF*) in the range of 2.2 - 4.2 with an average value of 3.1, and at the same time the ground-source heat pumps perform with *SPF* in the range of 3.0 - 5.4 with the average value of 4.3 [2]. The numbers indicated above are for newly built single-family dwellings. Furthermore, European standard on design of heat pump heating systems specifies the minimum and target values of heat pump performance, i.e. for air-source heat pump minimum and target *SPF* values are 2.3 and 2.8, respectively and 3.0 and 3.5 for the ground-source heat pumps with application for DHW and SH.

Thus, the study of air- and ground-source heat pump performance for DHW preparation under different climate conditions in Europe was carried out. The results demonstrate the scope of heat pump applicability for DHW preparation in European countries in regard to the standard [3].

2. Analysis of water heating heat pump system

Use of air source pump (ASHP) and ground source heat pump (GSHP) for DHW preparation has been analyzed for single family house with 4 persons with total hot water demand of 206 l at 45 °C. Analysis has been performed by means of simulations using TRNSYS software [4]. Analyzed system for DHW preparation consists of heat pump, hot water storage (TES), circulating pumps, controller and back-up heater (see Figure 1).

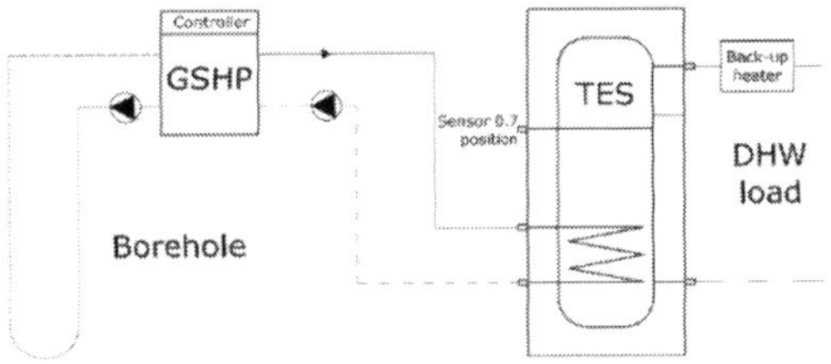

Fig. 1. Scheme of water heating ground-source heat pump system.

The heat pumps of both types have been sized to allow the hot water preparation and there were considered specific products available on the market. ASHP has the heat output 6.6 kW and $COP = 3.1$ at A2/W35 conditions, at the same time GSHP has a heat output 5.8 kW and $COP = 4.5$ at B0/W35 conditions. The performance of heat pumps has been modelled with use of bi-quadratic curve fits for the heat output and electric power input. The curve-fit parameters were obtained for commercially available heat pumps by least-square fitting from manufacturer's data [5] (see Figure 2). For the modelling of the heat pump performance, several restrictions have been imposed on heat pump operation, such as maximum

* Corresponding author: Yauheni.Kachalouski@fs.cvut.cz

https://doi.org/10.1051/e3sconf/202018203006

and minimum condenser and evaporator operating temperatures and minimal standby time between its start and stop (10 min).

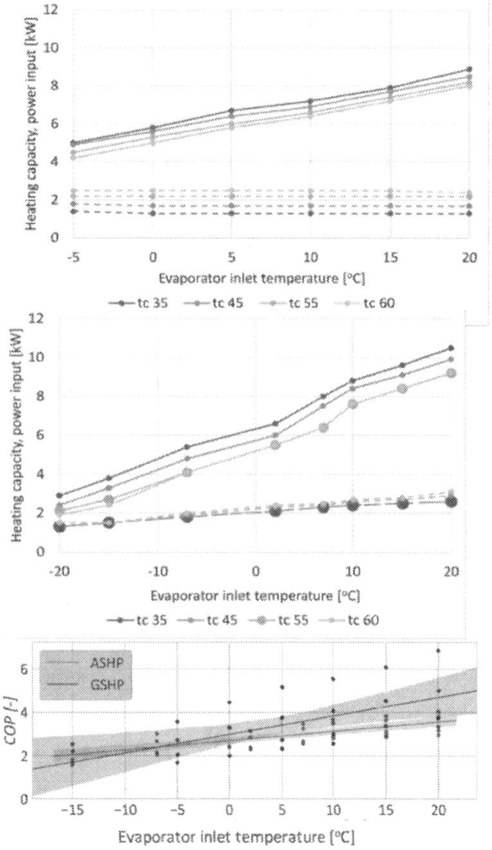

Fig. 2. Heat output, compressor power input and *COP* for air-source and ground-source heat pumps

The volume of the hot water storage is considered 200 l. To transfer the heat from the heat pump into water the immersed tube heat exchanger is applied. DHW tank has been modelled as the stratified fluid storage tank with internal heat exchanger known as type 340 [6]. The *UA*-value of heat exchanger is 640 W/K and it is considered as a function of temperature difference between fluid in the heat exchanger and in the storage. The heat loss of the DHW tank is characterized by overall UA-value of 1.58 W/K, distributed by 23 % for top heat loss, 70 % to wall heat loss and 7 % to bottom heat loss. The whole storage volume is divided into 30 completely mixed volume segments. The temperature sensor is placed at the 0.7 of relative height of the total tank height. Set-point temperature of sensor in the storage tank for hot water preparation control is 50 °C with a deadband of ± 2 K.

Ground-source heat pump is equipped with single U-tube borehole of 68 m depth. Here, to simplify the parametric analysis the sizing has been used universally for all cases. The model of borehole used in simulations is type 557 of TESS library for TRNSYS [7]. The ground and fill material thermal conductivity was set to 2 W/m.K.

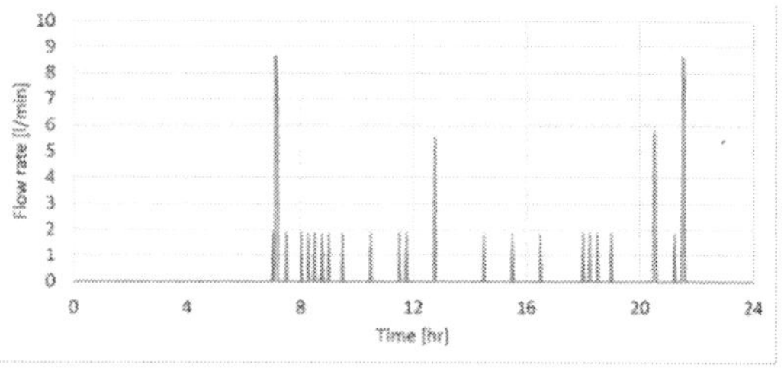

Fig. 3. DHW load profile adopted from M profile [8].

To model a realistic hot water load with morning/evening peaks and night load profile M has been taken from EU Commission Regulation 814/2013 on eco-design of water heaters [8]. Daily profile corresponding to the hot water consumption of single family of 4 persons is shown in Figure 3. However, the load profile M was adjusted from delivering temperature of 55 °C to 45 °C temperature of water intake keeping the same amount of energy consumed (see Figure 2). Reduction of the hot water temperature has been applied as more realistic requirement when using heat pumps in single-family houses.

The heat pump system has been considered in different climate locations throughout the Europe. Hourly climate data for each location has been obtained from meteorological database Meteonorm [4]. Data sets for 358 locations in the Europe (see Figure 4) have been selected for the analysis. The northernmost location is Barentsburg, Norway (GPS 78.1, 14.2), the southernmost location is Funchal, Portugal (GPS 32.6, -16.9). Cold water mains temperature has been derived by using the following algorithm adopted from [9]:

$$t_{mains} = (\bar{t}_{amb} + off) + r\left(\frac{t_{amb,max}}{2}\right)\sin\left(\frac{360}{365}(day - 15 - lag) - 90\right) \quad (1)$$

where is

$\bar{t}_{amb}$, $t_{amb,max}$	average and maximum ambient temperature [°C];
off, r, lag	fitting coefficients [-], offset, ratio and lag values were obtained by fitting data compiled by Abrams and Shedd and Sandia Labs [9];
day	current day of simulation [day].

https://doi.org/10.1051/e3sconf/202018203006

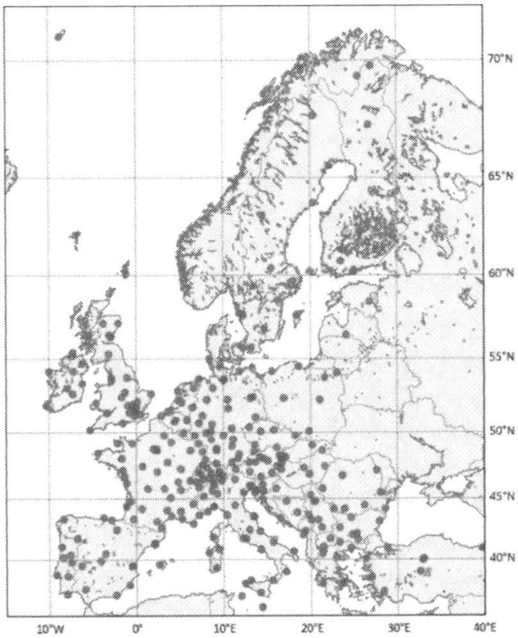

Fig. 4. Locations considered in the simulations.

The operation conditions of hot water preparation by air source heat pump: ambient air temperature and cold water temperature are influenced not only by geographical latitude but also altitude (see Figure 5). Data of high altitude around latitude value of 48° is for locations in Alps mountains. The fitting coefficients proposed by Abrams and Shedd in Equation 1 lead the water mains temperatures decrease below 0°C in high latitudes, hence it was suggested to limit water mains temperature down to 4°C in these latitudes, which is seen in Figure 5.

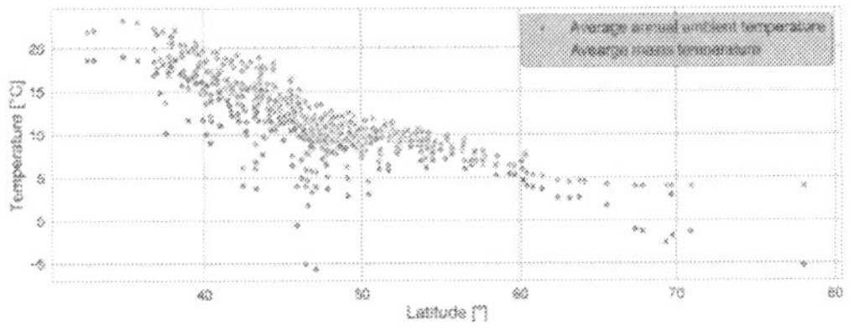

Fig. 5. Average annual ambient and mains temperature.

The extensive number of numerical simulations were performed in TRNSYS software [4] which was controlled over the Python script. This approach allows to generate newly compounded TRNSYS input file (*.dck-file) containing all the boundary conditions for each simulation.

3. Results

Heat demand for hot water preparation differs in different locations (Figure 6) due to the water mains temperature differs throughout the year (Figure 5) influencing DHW energy delivered to the consumer. Thus, heat demand increases with latitude when moving to the north or with the altitude of the location. It is about double in the mountain areas or northern part of Europe than it is in the southern Europe.

Moving from south to north increasing latitude of the location the total system electrical energy consumption (which includes electrical consumption of heat pump, circulation pump and back-up heater) increases. Again, the spike between 45° and 50° latitude indicates high altitude of given sites in mountains characterized by significantly higher heat demand for hot water preparation (see Figure 6) and extreme conditions for heat pump operation. System total electrical consumption in this mountain area is thus approximately 1.5 times higher than it is in the same latitude location but with low altitude (1500 kWh versus 1000 kWh).

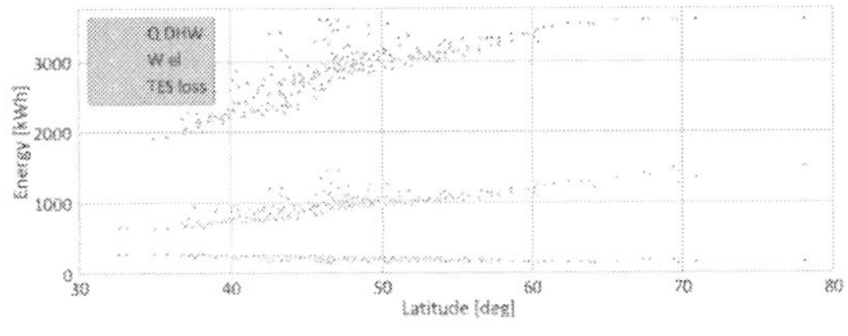

Fig. 6. DHW heat demand, system electrical energy consumption and TES heat loss for different locations (latitude) for ASHP.

In general air-source and ground-source heat pump system simulations resulted in similar system performance for given location. Figure 6 provides simulations results of DHW heat demand, system electricity consumption and storage tank heat loss for air-source heat pump only. Due to these values are barely distinguishable from those obtained for the ground-source heat pump, to investigate key difference in performance of two types of heat pumps the *SPF* values were evaluate alone (Figure 7).

Significant correlation was identified between system performance and the boundary conditions the heat pump was exposed to. By exploiting overall system seasonal performance factor (*SPF*) the heat pump system can be evaluated as follows:

$$SPF = \frac{Q_{DHW}}{W_{el,hp} + W_{el,aux} + W_{el,bu}} \quad (2)$$

where is

Q_{DHW} heat delivered to hot water load [kWh];
$W_{el,hp}$ electricity consumed by heat pump [kWh];
$W_{el,aux}$ electricity consumed by circulating pump [kWh];
$W_{el,bu}$ electricity consumed by back-up heater [kWh].

The efficiency of heat pump described as coefficient of performance *COP*. The annual *COP* has been calculated according to the Equation (2) as follows:

$$COP = \frac{Q_c}{W_{el,hp}} \quad (3)$$

where is

Q_c condenser heat output [kWh];
$W_{el,hp}$ electricity consumed by heat pump [kWh].

https://doi.org/10.1051/e3sconf/202018203006

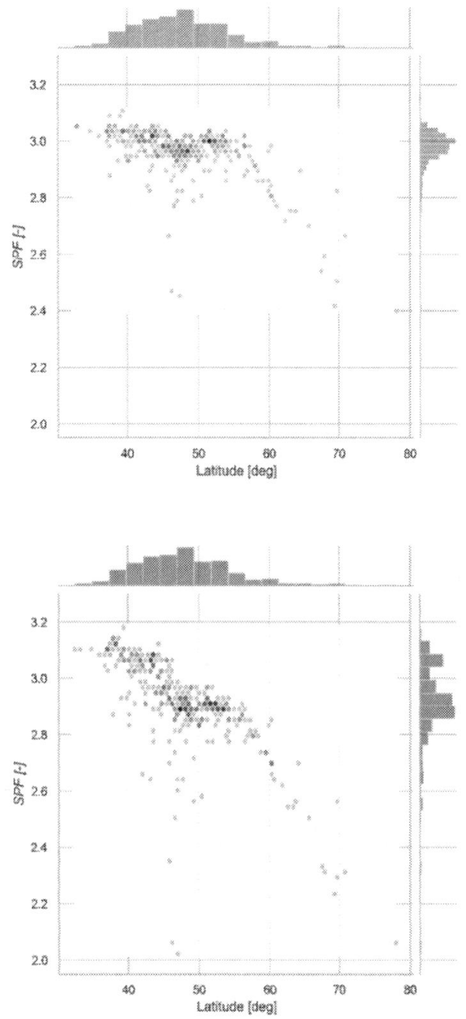

Fig. 7. *SPF* of heat pump system in different locations: top – ASHP, bottom – GSHP.

Figure 7 reflects that seasonal performance factor *SPF* of ASHP at considered locations is close to normal distribution with peak in value 3.0. On the other hand, GSHP has two peaks for *SPF* near the 3.1 and 2.9 values. Also, one can see that GSHP outperforms ASHP in warm climate (low latitudes) but has a bit worse efficiency in mid- and high latitudes (mid- and cold climate). This fact can be good explained from the Figure 3 that reveals the heat pump coefficients of performance, and the GSHP has a steeper global regression line of *COP*.

As it is known the heat pump performance is dependent on the condensing and evaporating temperatures, hence in case of air source heat pump technology on the condenser supply temperature (which is influenced by water mains temperature) and on the ambient temperature. Thus, the only parameters that have a key influence on the heat pump *COP* and *SPF* are the ambient temperature and water mains temperature.

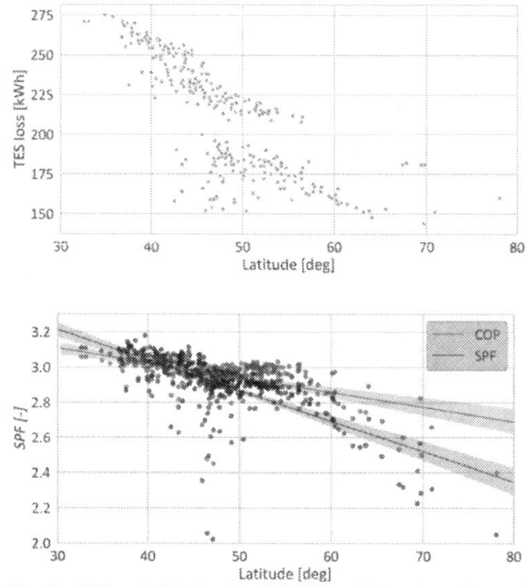

Fig. 8. *COP* and *SPF* (top) and DHW tank heat loss (bottom) of ASHP in different locations

The heat pump system efficiency decreases due to the operation under low inlet temperatures to the evaporator in high latitudes and/or altitudes. System *SPF* in all locations has less obvious trend than the trend for *COP* of the heat pump, but it is also decreasing with the latitude (see Figure 8). It is given by the fact that the heat loss of hot water storage tank is covered by heat pump and thus included in surplus electricity consumption and reflected in worsening the system parameter *SPF*, while *COP* characterizes only the efficiency of the heat pump alone. The heat loss from the hot water storage tank to the surroundings is also varying with the location, it is higher in warmer climate than it is in colder climate. It seems surprising at least for first sight, however it is logic and it is followed by detailed explanation.

The heat loss of storage tank to the surroundings occurs due to temperature difference between the storage tank volume and the surroundings (technical room where storage tank is placed), as follows:

$$\dot{Q}_{loss} = UA(t_{ST} - \bar{t}_r) \tag{4}$$

where is

U	overall heat transfer coefficient of the storage tank [W/(m²K)];
A	surface area of the storage tank envelope [m²];
t_{ST}	mean storage tank temperature [°C];
$\bar{t}_r$	room temperature [°C], for the simulations considered to be constant and equal to 20 °C.

As one can see from the Equation 3 the higher the temperature difference between DHW tank and the surroundings, the higher heat loss rate under constant UA-value. Considering a stratification behavior of the DHW tank and that the temperature sensor is installed at the top part of the tank (at 70 % of the tank height), the bottom part of the DHW tank after finished charging cycle is usually filled with the cold water and it is not mixed with the top part, due to the thermocline. Thus, the top part of DHW tank is heated up to the desired set temperature 50 °C, however the bottom part remains at low temperature close to the temperature of cold water. Since the set temperature of the DHW tank is the same in all the simulations, temperature of the cold water mains has the huge impact on the mean tank temperature. Therefore, the lower the water mains temperature is the lower the mean tank temperature t_{ST} finally will be. Higher temperature of water mains in low latitude locations then results in higher heat loss of DHW system (see Figure 8).

To show the influence of water mains temperature on heat loss and further to seasonal efficiency expressed by *SPF*, three locations have been selected representing warm climate (Athens, Greece), medium climate (Strasbourg, France) and cold climate (Helsinki, Finland).

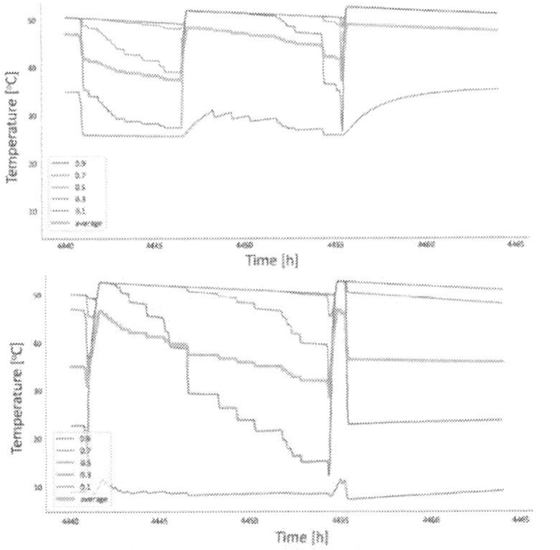

Fig. 9 System one-day simulation in Athens (upper) and Helsinki (lower) climate.

Simulation of one-day period is compared for significantly distinct locations Athens and Helsinki. Temperature in 5 positions of DHW tank relative height (0.9 / 0.7 / 0.5 / 0.3 / 0.1) has been monitored during the simulations. As it can be seen from Figure 10 the heat pump operates before large tapping in the evening (21 h) for cold climate causing mean DHW tank temperature to drop after this tapping occurred, though monitoring temperature remains high enough for the heat pump controller not to start DHW tank charging again. Since there is no tapping during the night time the mean DHW tank temperature is lower in comparison with the warmer climate (Figure 10), where the heat pump operates even after last tapping in the day keeping the DHW tank charged and at high temperature.

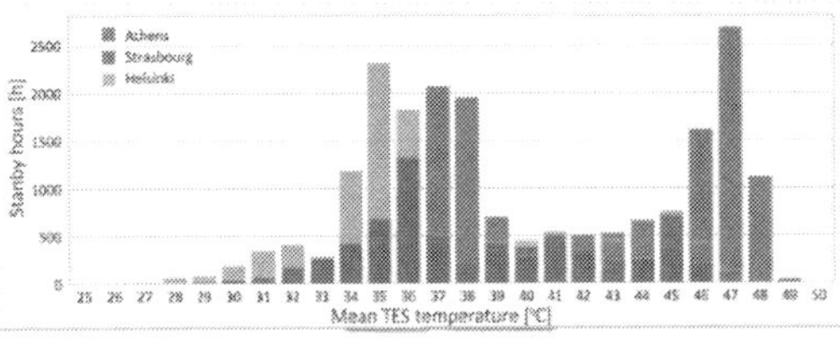

Fig. 10 DHW tank standby hours versus different mean storage temperatures in three climate locations.

As it can be seen from Figure 9, mean water temperature in the tank in cold climate is lower compared with southern climate, this consequently causes lower rate of heat loss to the surroundings (20 °C) in colder climate (Helsinki) compared to warmer climate (Athens).

Further, for the Helsinki climate simulation (Figure 10) the bottom part of storage tank is colder than surroundings leading to the heat flow direction to be from surrounding to the tank, i.e. bottom part of the storage tank obtains heat gains. This phenomenon has in its turn an influence on DHW tank heat loss at given location. Figure 10 also reveals a strong temperature stratification in case of colder water inlet to the DHW tank, which is no longer take place in warm climate (sensor 0.1).

The observed bias of TES loss can be satisfactorily explained with Figure 10 presenting TES standby hours at each mean temperature in three climates. Selected system setup (component parameters, control settings, etc.) often results in higher mean temperature of the storage in warm climate (Athens) compared to cold climate (Helsinki).

As discussed and concluded above, GSHP has a steeper regression line of *COP* (see Figure 3), nevertheless on the other hand system *SPF* is greater for ASHP than GSHP (Figure 7). To shed light on this discrepancy Figure 11 compares GSHP evaporator inlet temperatures for three climates. If one compares average ambient air temperature, i.e. evaporator inlet temperature for ASHP, with average borehole outlet temperatures, i.e. evaporator inlet temperature for GSHP, for the same location GSHP outperforms ASHP only in warm climate, despite the fact the ambient air temperature Athens is 17.6 °C, while outlet of borehole keeps at level of 7.5 °C (Figure 11). Temperature values for given locations are given in the Table 1.

Table 1. Average evaporator inlet temperatures for ASHP and GSHP for three climate locations.

Heat pump technology / location	**Average ambient air temperature (ASHP evaporator inlet)**	**Average borehole outlet temperature (GSHP evaporator inlet)**
Athens	17.6	7.5
Strasbourg	9.8	1.5
Helsinki	4.5	-2.5

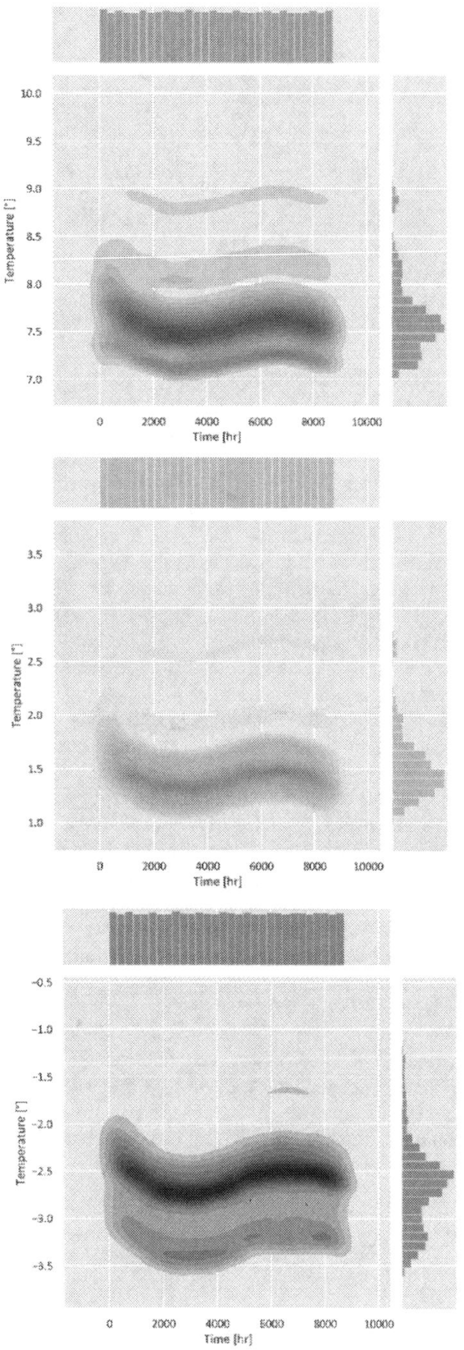

Fig. 11 GSHP evaporator inlet temperature: top – Athens, center – Strasbourg, bottom – Helsinki climate.

4 Conclusions

The system analysis shows that heat pump performance for hot water preparation is strongly dependent on the climate, i.e. ambient temperature and temperature of the water mains. The highest SPF reached by the ASHP during this study was around 3.1, the lowest – 2.4, however the GSHP simulations resulted in *SPF* of 3.2 - 2.0 range. Average *SPF* values were found to be 3.0 and 2.9 for ASHP and GSHP, respectively. Average values are in good correspondence with target values of the standard [3] which were given in the introduction.

Though this study was limited to the same setup for all locations it demonstrated that such system (air- and ground-source performance does not drop lower than minimum value of 2.3, mainly due to lower DHW water intake temperature of 45 °C versus widely used 55 °C). It is worth to mention that the results obtained are in good compliance with the measurements of real installations [2].

The authors would like to acknowledge the financial support received for this research from NPU (National Sustainability Program I, project No. LO1605).

References

1. Author, European Heat Pump Market and Statistics Report 2015 Executive Summary. pp. 7–10, EHPA, 2015.
2. M. Miara, "Efficiency of Heat Pumps in Real Operating Conditions – Results of three Monitoring Campaigns in Germany," REHVA Eur. HVAC J., vol. 51, pp. 5–12, 2014.
3. EN 15450 Heating systems in buildings – Design of heat pump heating systems, CEN, 2008.
4. U. of W.--M. S. E. Laboratory, TRNSYS, a transient simulation program. Madison, Wis.: The Laboratory, 1975., 1975.
5. E. Jones, T. Oliphant, P. Peterson, and others, "{SciPy}: Open source scientific tools for {Python}."
6. H. Drück and D.-I. H. habil Müller-Steinhagen Pfaffenwaldring, "MULTIPORT Store -Model for TRNSYS Stratified fluid storage tank with four internal heat exchangers, ten connections for direct charge and discharge and an internal electrical heater Type 340," p. 27, 2006.
7. G. Hellström, "Duct Ground Heat Storage Model, Manual for Computer Code," Energy, no. March, 1989.
8. E. U. C. Regulation, EU Commission Regulation 814/2013 implementing Directive 2009/125/EC of the European Parliament and of the Council with regard to ecodesign requirements for water heaters and hot water storage tanks, 2013, Brussel, 2013.
9. D. W. Abrams and A. C. Shedd, "Effect of seasonal changes in use patterns and cold inlet water temperature on water-heating loads," ASHRAE Trans., vol. 102, pp. 1038–1053, 1996.

 https://doi.org/10.1051/e3sconf/202018203007

Combining machine learning algorithms and an incremental capacity analysis on 18650 cell under different cycling temperature and SOC range

John Lai*, *David* Chao, *Alvin* Wu, *Carl* Wang

Underwriters Laboratories Taiwan Co. Ltd., No. 260, Daye Rd., Beitou Dist., Taipei City 112, Taiwan

Abstract. A novel way to apply machine learning algorithms on the incremental capacity analysis (dQ/dV) is developed to identify battery cycling conditions under different temperatures and working SOC ranges. Batteries are cycled under each combination of temperatures (-10°C, 25°C, 60°C) and SOC ranges (0-10%, 25-75%, 90-100%, 0-100%) up to 60 equivalent cycles. The discharge data is transformed into dQ/dV-V curve and its features of the peaks and valleys are further taken for machine learning. Both supervised and unsupervised machine learning algorithms (PCA and LDA) are applied to classify batteries in terms of temperature or SOC range. The results reveal that batteries cycled under different temperatures can be identified separately regardless of the working SOC range. When splitting 60 samples with a ratio of training set equals to 0.85, the remaining test set gives an identification accuracy of 89% in temperature and 67% in working SOC range.

1 Introduction

Rechargeable batteries have been widely used in our daily life, for example, from a stationary electrical storage system connected to an electric grid to portable devices such as smartphones and laptops. Among several types of rechargeable batteries, the lithium-ion battery is the dominant one with high energy density and working voltage. Nowadays, lithium-ion batteries can be easily found on 3C products and electric vehicles. The real battery usage in these applications suffers degradation from different temperatures and working SOC ranges due to environmental temperature, device design, and user's habit that makes life evaluation difficult [1-3]. There have been massive researches on battery aging behavior under different temperature and working SOC ranges owing to their significant impact on battery degradation mechanisms, which applies huge uncertainty and variation in battery lifespan and safety [4-8]. Thus, there has been numerous inspection methods developed to evaluate the cell aging status. One of the well-known methods is called incremental capacity (dQ/dV) analysis, which transfers charging/discharging data into dQ/dV-V curves followed by observing peak shifts or amplitude changes [9-11]. Although many tests have been conducted and shifts of peaks have been explained in previous studies, an easy, general and accurate way to identify or predict battery cycling conditions remains unrevealed. Recently, machine learning as well as deep learning have privileged in all industry including lithium-ion battery to address complex problems by processing classification or providing more insight from only data. It has been largely studied in state of health (SOH) and remaining useful life (RUL) prediction [12-13].Hence, this study tries to apply machine learning algorithms on features extracted from dQ/dV-V curve to investigate changing directions of these features under different cycling temperatures and working SOC ranges in pursuit of the identification of battery aging status.

2 Experimental procedure

Commercial Panasonic 18650 cell, 3350 mAh, was selected for the test. All cells were conducted 3-cycle conditioning before experiments to check cell consistency, these conditioned samples were called fresh cells hereafter. Each cell corresponded to one of combinations composed of different temperature (-10°C, 25°C, 60°C) and working SOC ranges (0-10%, 25-75%, 90-100%, 0-100%). The cells were cycled by CC discharge mode and CC-CV charge mode with 0.2C within a voltage range between 2.5V and 4.2V and a cutoff current of 0.02C. For 0-10% SOC, cells were fully discharged before charge/discharge cycles. For 25-75% SOC, cells were discharged to 25% SOC and then followed by charge/discharge cycles. For 90-100% and 0-100% SOC, cells were directly cycled from a fully charged state. To ensure all cells are cycled under proper SOC ranges, the maximum capacity at each cycling temperature was measured as a baseline to estimate charged/discharge time. The charge/discharge time and cycles for 15 equivalent cycles under different temperature and SOC ranges are summarized in Table. 1. After every 15 equivalent cycles, cells were conducted a 2-cycle retention capacity test according to battery specification,

* Corresponding author: john.lai@ul.com

 https://doi.org/10.1051/e3sconf/202018203007

0.5C charge with a cutoff current equals to 0.02C and 0.2C discharge, followed by a 24-hour rest to record OCV drop. The 2nd discharge data in the retention capacity test was further used for incremental capacity (dQ/dV) analysis and machine learning.

Table 1. Cycling parameters for 15 equivalent cycles under different temperatures and SOC ranges.

SOC range	-10 °C		25 °C		60 °C	
	C/D time	Cycles	C/D time	Cycles	C/D time	Cycles
0-10%	1504s	150	1781s	150	1841s	150
25-75%	7520s	30	8905s	30	9205s	30
90-100%	1504s	150	1781s	150	1841s	150
0-100%	*	15	*	15	*	15

*Cells stop charge/discharge when reaching termination conditions.

3 Experimental procedure

3.1 Battery cycling data

The discharge capacity and 24-hour OCV drop of fresh cells (3-cycle conditioning) and cycled cells (15, 30, 45, 60 equivalent cycles) are summarized in Fig. 1 and Fig. 2 labeled with temperature and working SOC range separately. In Fig. 1, cells cycled under -10 °C and 60 °C show lower retention capacity but cannot be separated clearly, and cells cycled at 25 °C shows higher capacity but still partially mix with cells cycled at 60 °C. Some of the cells cycled under -10 °C and 60 °C have an obvious higher OCV drop after 24-hour rest but the remaining cells have lower OCV drop than fresh cells and are mixed with cells cycled at 25 °C. In Fig. 2, 0-10% SOC cycled cell has slightly higher retention capacity while other SOC ranges are mixed together. Similarly, most of the cells have OCV drops lower than fresh cells and are inseparable. Based on the result, the aging effect caused by cycling temperature or working SOC range factors is not easy to be identified by simply retention capacity and 24-hour OCV drop data.

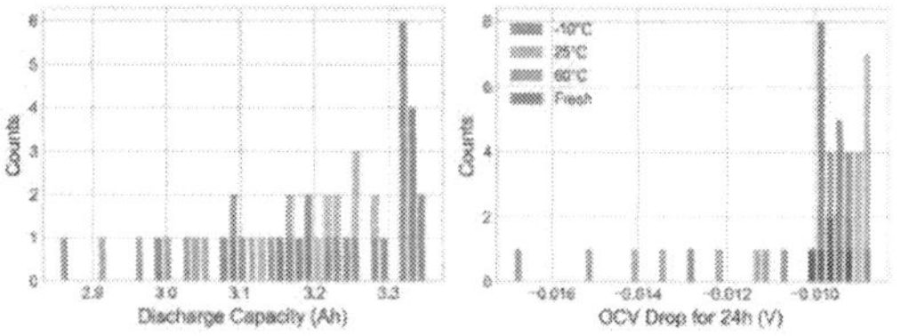

Fig. 1. The discharge capacity and 24-hour OCV drop of fresh and cycles under different cycling temperature.

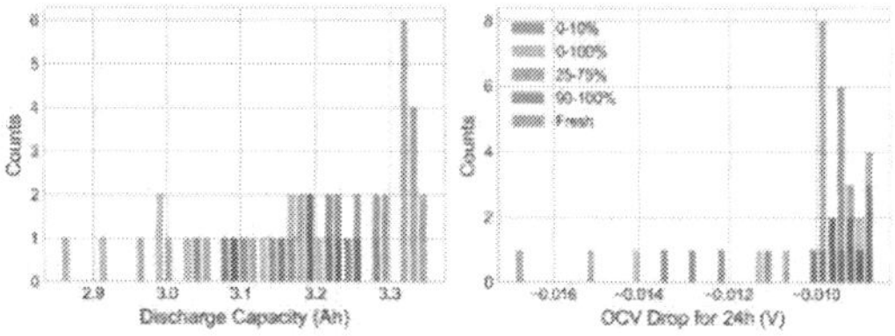

Fig. 2. The discharge capacity and 24-hour OCV drop of fresh and cycles under different cycling SOC range.

3.2 Features extracted from dQ/dV-V curve

The incremental capacity analysis and feature extraction process applied on 2nd discharge data in the retention capacity test are summarized in Fig. 3(a) and follows the steps: (i) slice original discharge data into multiple segments with a voltage interval of 0.011V from 4.2V to 2.5V; (ii) calculate the capacity difference (dQ) in each segment and divided by 0.011V (dV) to obtain dQ/dV value; (iii) calculate mean voltage of each segment; (iv) take dQ/dV as y-axis and mean voltage as x-axis to plot dQ/dV-V curve. The dQ/dV-V curves of fresh cells (red solid line) and cycled cells (black dash line) are shown in Fig. 3(b). The study of dQ/dV-V curve generally focuses on the positions of peaks and valleys because they represent a phase transformation in cathode or intercalation of lithium into a graphite anode that is highly-related to cell aging behavior. Hence, four peaks (P1, P2, P3, P4) and three valleys (V1, V2, V3) are marked in Fig. 3(b), and their x-values and y-values are used as features for algorithms to make grouping or classification later in the sections 2.4 and 2.5. For example, the x-value and y-value of peak4 (P4) are marked as VP4 and IP4 respectively. From the dQ/dV-V curves, the fresh cells with uniform curves usually indicate the cell quality is consistent and it serves as a baseline to observe how curve shifts after cells are cycled. After being aged, cells show the following trends: peak1 (P1) shifts to an upper-left place; peak2 (P2), peak3 (P3), peak4 (P4) and valley2 (V2) shift to a lower place; valley1 (V1) and valley3 (V3) shift to both an upper and a lower place. It should be noticed that the curves near peak3 (P3), valley3 (V3), and peak4 (P4) have large variation without a consistent trend, which might be caused by different cycling conditions.

https://doi.org/10.1051/e3sconf/202018203007

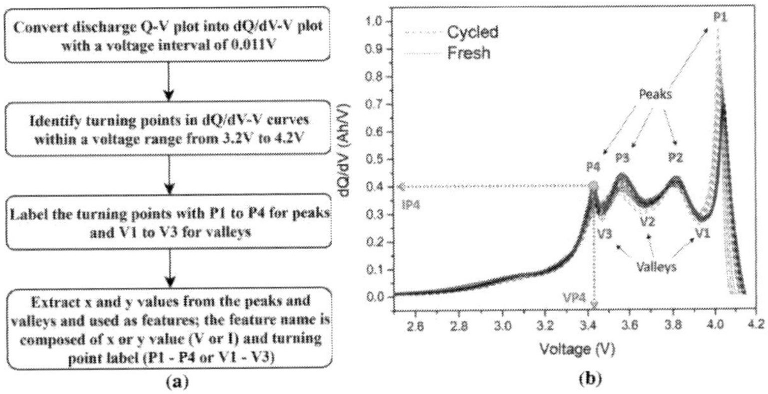

Fig. 3. (**a**) the flowchart of data processing from Q-V plot into dQ/dV-V plot and feature extraction from dQ/dV-V curve; (**b**) the dQ/dV-V curves of fresh (red solid line) and cycled cells (black dash line).

The features extracted from the dQ/dV-V curve are further standardized in order to improve algorithm results. To be more precise, the standardization rescales data with a mean of 0 and a standard deviation of 1, and it is widely used before machine learning algorithms to avoid being governed by a particular feature having a broad range of values. The normalized features labeled with cycling temperature and SOC range are shown in Figs. 4-7. For different cycling temperatures, 60 °C cycled cells have higher VV3 and IP1 than -10 °C cycled cells, but lower IV3 than -10 °C cycled cells. Among all y-value (I) features, 25 °C cycled cells locate in the middle of the distribution. In contrast, for different cycling SOC ranges, only 0-10% SOC cycled cells show higher IV2 and lower IV3 clearly among all cells. The remaining SOC ranges don't provide any obvious trend. Overall, the trends mentioned above are unable to separate cycled cells under different temperatures or working SOC ranges. In next section, PCA will be applied to reduce the data dimension to visualize data distribution.

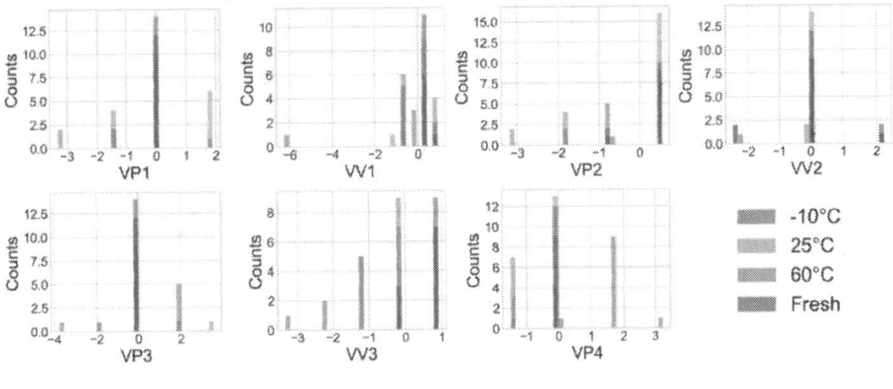

Fig. 4. Normalized data of x-value (V) of each peak (P1-P4) and valley (V1-V3) in dQ/dV-V curve labeled with cycling temperature.

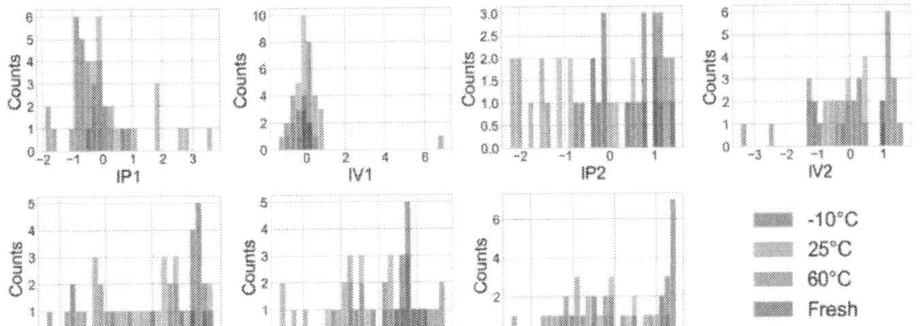

Fig. 5. Normalized data of y-value (I) of each peak (P1-P4) and valley (V1-V3) in dQ/dV-V curve labeled with cycling temperature.

https://doi.org/10.1051/e3sconf/202018203007

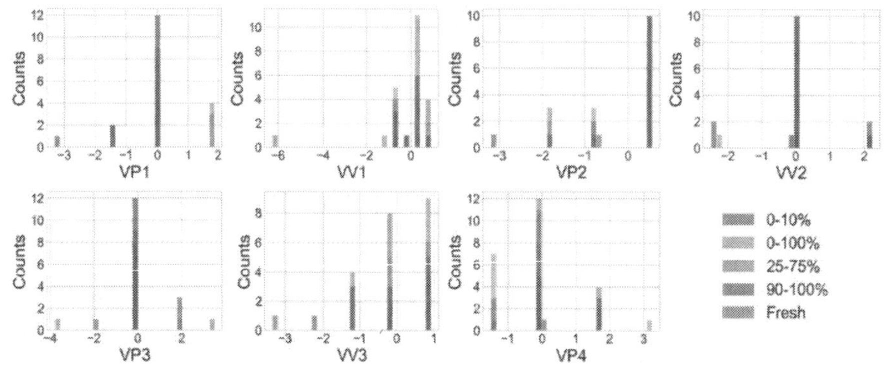

Fig. 6. Normalized data of x-value (V) of each peak (P1-P4) and valley (V1-V3) in dQ/dV-V curve labeled with cycling SOC range.

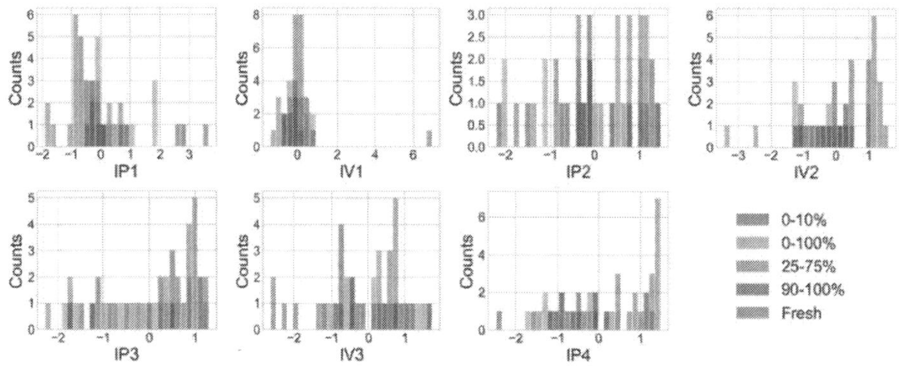

Fig. 7. Normalized data of y-value (I) of each peak (P1-P4) and valley (V1-V3) in dQ/dV-V curve labeled with cycling temperature.

3.3 Unsupervised algorithm (PCA) analysis

Principle component analysis (PCA) is an unsupervised algorithm used to keep maximum data variation when reducing data dimensions to provide a visualized data distribution. The PCA plots of cells cycled under different temperatures and working SOC ranges are shown in Fig. 8 (a) and (b) respectively. In Fig. 8 (a), the fresh cells have the most narrow distribution, while cells cycled under -10 °C and 60 °C show the widest distribution among all. It indicates that cells cycled under -10 °C and 60 °C will cause much more variation in features than those cycled at 25 °C. Besides, a slight grouping can be observed along PC2-axis but there are still around half of the cells mixed together. In contrast, the PCA result of different SOC ranges provides not much information but wide and mixed distributions among all cycled cells. It can be inferred that the working SOC range effect may be suppressed by the temperature factor in feature variation.

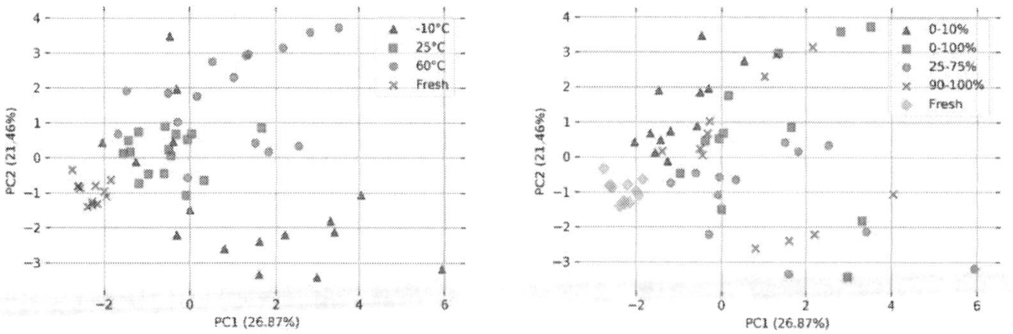

(a) (b)

Fig. 8. PCA plots of cells cycled under different: (a) temperature (b) SOC range.

https://doi.org/10.1051/e3sconf/202018203007

When looking into the cumulative explained variance chart shown in Fig. 9, the first two principal components only accounts for 48% explained variance. It indicates that multiple features cause similar variance so that even after rotating the principal component such as PC1 and PC2, the original variance is still unable to be effectively explained in a 2-D projected plane. Thus, a supervised algorithm is applied in the next section to obtain more explained variation and achieve the classification of cycling conditions.

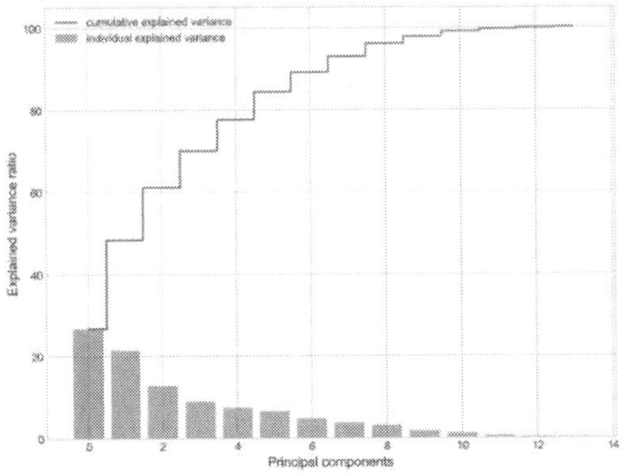

Fig. 9. PCA cumulative explained variance along with the number of principal components.

3.4 Supervised algorithm (LDA) analysis

Linear discriminant analysis (LDA) is a supervised algorithm used to maximize the gap between groups but minimize internal differences within a group. The LDA plots of different cycled temperatures and SOC ranges are given in Fig. 10 (a) and (b) respectively. According to the evaluation test in a test set ratio of 0.15, different cycling temperature gives an accuracy of 89%, while different SOC range only has an accuracy of 67%. In Fig. 10 (a), cells cycled under different temperatures can be obviously separated into three blocks along LD1-axis. It indicates that the LD1-axis is highly related to temperature and its values of 0.2 and -0.2 can be used as a simple way to identify cycling temperature. For example, cycled cells that have LD1 higher than 0.2 can be identified as 60 °C cycled cells. It should be noticed that this way of identifying cycling temperature applies to all working SOC ranges. On the other hand, cells cycled under different working SOC ranges are unable to be thoroughly categorized as four blocks. If the 90-100% SOC range is removed, the remaining SOC ranges can be identified along LD1-axis with specific values of 0.3 and -0.2. To understand the relationship between the LD1-axis and the dQ/dV-V curve, the eigenvalues are further discussed in the next section.

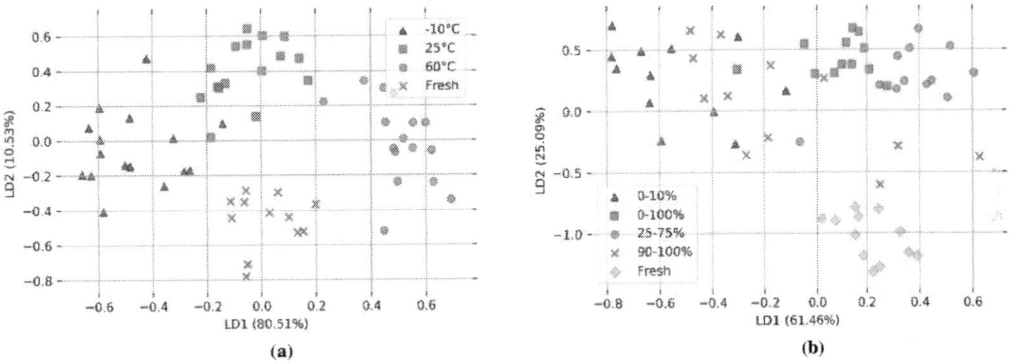

(a) (b)

Fig. 10. LDA plots of cells cycled under different: (a) temperature (b) SOC range.

3.5 Insight into dQ/dV-V curve from eigenvalues

In both PCA and LDA, the axes are also called eigenvectors, which are linear vectors composed of each feature multiplied by each corresponding eigenvalues. The higher the eigenvalue, the more important the corresponding feature in contributing explained variation. Generally in LDA, when a feature has a high eigenvalue, it represents the feature has more power to classify samples according to their labels. The eigenvalues of PCA and LDA (temperature and working SOC range) algorithms are summarized in Table. 2. From the LDA (temperature) result, the VP4, IP3, IV3, and IP4 features have high eigenvalues marked in bold in LD1-axis, and LD1-axis is shown capable of classifying cycling temperature in Fig. 10 (a). Thus these features are highly related to cycling temperature. In addition, these features correspond to peak3 (P3), valley (V3), peak4 (P4) in dQ/dV-V curve, shown in Fig. 3(b), ranging from 3.4V to 3.6V. It indicates this specific voltage range in dQ/dV-V curve is highly temperature-dependent and its shifting behavior can be used to identify battery cycling temperature. On the other hand, we can observe the LD1 eigenvalues in LDA (SOC range) and those in LDA (temperature) are nearly the same. It indicates most of the explained variation in LDA (SOC range) might be caused by temperature factor. In other words, the cycling temperature has a more dominant impact than the working SOC range in the shifting behavior of dQ/dV-V curves.

Table 2. Eigenvalues of PCA and LDA algorithms.

Feature	PCA (Temperature & SOC range)		LDA (Temperature)		LDA (SOC range)	
	PC1 (26.87%)	PC2 (21.46%)	LD1 (80.51%)	LD2 (10.53%)	LD1 (61.46%)	LD2 (25.09%)
VP1	0.20	0.20	0.16	0.13	0.19	0.19
VV1	0.12	0.16	0.04	-0.07	0.06	-0.05
VP2	0.29	0.05	-0.02	0.04	-0.09	0.10
VV2	0.11	-0.21	0.04	0.09	-0.01	0.08
VP3	0.10	-0.22	-0.06	-0.08	-0.06	0.04
VV3	0.28	-0.18	0.04	0.02	-0.13	0.17
VP4	-0.12	0.44	**-0.24**	0.23	**-0.28**	0.40
IP1	-0.13	-0.46	0.20	0.20	0.09	0.26
IV1	0.10	-0.30	0.05	-0.10	0.00	-0.05
IP2	0.14	0.36	0.16	-0.14	0.07	-0.39
IV2	0.50	-0.04	-0.17	-0.31	0.00	-0.43
IP3	0.49	-0.07	**-0.54**	0.75	**-0.68**	0.47
IV3	0.06	0.41	**-0.25**	0.04	**0.28**	0.00
IP4	0.46	0.05	**0.67**	-0.42	**0.55**	-0.37

4 Conclusion

The incremental analysis (dQ/dV) analysis is an effective way in diagnosing the battery aging behavior and a supervised algorithm is useful to classify cycling temperature and provide insight into the shifting behavior of dQ/dV-V curves. From the observation of retention capacity, 24-hour OCV drop, peak and valley positions in dQ/dV-V curve, some insignificant trends can be found but too complicated and insufficient to identify cycling temperature or working SOC range. By applying a supervised LDA algorithm, cells cycled under different temperatures (-10°C, 25°C, and 60°C) can be well separated into three blocks in a 2-D projected plane. According to an evaluation test, the identified accuracy reaches 89% in a test set ratio of 0.15. Further study on eigenvalues of the LD1-axis reveals that the effect of cycling temperature mainly reflects on a specific voltage range from 3.4V to 3.6V in the dQ/dV-V curve. On the other hand, although cells cycled under different SOC ranges cannot be well identified, the eigenvalue information shows most data variance is caused by the cycling temperature factor. In other words, the cycling temperature is more dominant than the working SOC range in the shifting behavior of dQ/dV-V curve.

References

1. Xiaopeng Tang, Yujie Wang, Changfu Zou, Ke Yao, Yongxiao Xia, Furong Gao. A novel framework for Lithium-ion battery modeling considering uncertainties of temperature and aging. Energy Convers., 2019; 180:162-170.
2. Anthony Barré, Benjamin Deguilhem, Sébastien Grolleau, Mathias Gérard, Frédéric Suard, Delphine Riu. A review on lithium-ion battery ageing mechanisms and estimations for automotive applications. Journal of Power Sources, 2013; 241:680-689.
3. Christoph R. Birkl, Matthew R. Roberts, Euan McTurk, Peter G. Bruce, David A. Howey. Degradation diagnostics for lithium ion cells. Journal of Power Sources, 2017; 341:373-386.
4. Feng Leng, Cher Ming Tan, Michael Pecht. Effect of temperature on the aging rate of Li ion battery operating above room temperature. Sci. Rep.,5, 12967; doi: 10.1038/srep12967 (2015).
5. Naixing Yang, Xiongwen Zhang, BinBin Shang, Guojun Li. Unbalanced discharging and aging due to temperature differences among the cells in a lithium-ion battery pack with parallel combination. Journal of Power Sources, 2016; 306:733-741.
6. Christoph R. Birkl, Matthew R. Roberts, Euan McTurk, Peter G. Bruce, David A. Howey. Degradation diagnostics for lithium ion cells. Journal of Power Sources, 2017; 341:373-386.
7. Issam Baghdadi, Olivier Briat, Jean-Yves Delétage, Philippe Gyan, Jean-Michel Vinassa. Lithium battery aging model based on Dakin's degradation approach. Journal of Power Sources, 2016; 325:273-285.
8. Yang Gao, Jiuchun Jiang, Caiping Zhang, Weige Zhang, Yan Jiang. Aging mechanisms under different state-of-charge ranges and the multi-indicators system of state-of-health for lithium-ion battery with Li(NiMnCo)O2 cathode. Journal of Power Sources, 2018; 400:641-651.
9. Matthiue Dubarry, Cyril Truchot, Boryann Liaw. Mechanistic diagnosis and prognosis on Li-ion battery degradation. Presented at: 2013 IBA.
10. Linjing Zhang, Jiuchun Jiang, Weige Zhang. Capacity Decay Mechanism of the LCO + NMC532/Graphite Cells Combined with Post-Mortem Technique. Energies. 10(8), 1147; doi: 10.3390/en10081147 (2017).

11. Tiphaine Plattard, Nathalie Barnel, Loïc Assaud, Sylvain Franger, Jean-Marc Duffault. Combining a Fatigue Model and an Incremental Capacity Analysis on a Commercial NMC/Graphite Cell under Constant Current Cycling with and without Calendar Aging. Batteries., 5(1), 36; doi: 10.3390/batteries5010036 (2019).
12. Adnan Nuhic, Tarik Terzimehic, Thomas Soczka-Guth, Michael Buchholz, Klaus Dietmayer. Health diagnosis and remaining useful life prognostics of lithium-ion batteries using data-driven methods. Journal of Power Sources, 2013; 239:680-688.
13. Xiaosong Hu, Shengbo Eben Li, Yalian Yang. Advanced Machine Learning Approach for Lithium-Ion Battery State Estimation in Electric Vehicles. IEEE Transactions on Transportation Electrification, 2016; 2:140-149.

AUTHOR INDEX